CHEMISTRY RESEARCH AND APPLICATIONS

PREFERENTIAL SOLVATION AND HYDRATION OF PROTEINS IN WATER-ORGANIC MIXTURES

TWO SIDES OF ONE COIN

CHEMISTRY RESEARCH AND APPLICATIONS

Additional books and e-books in this series can be found on Nova's website under the Series tab.

CHEMISTRY RESEARCH AND APPLICATIONS

PREFERENTIAL SOLVATION AND HYDRATION OF PROTEINS IN WATER-ORGANIC MIXTURES

TWO SIDES OF ONE COIN

VLADIMIR A. SIROTKIN
WITH CONTRIBUTION FROM
ALEXANDRA A. KUCHIERSKAYA

Copyright © 2019 by Nova Science Publishers, Inc.

All rights reserved. No part of this book may be reproduced, stored in a retrieval system or transmitted in any form or by any means: electronic, electrostatic, magnetic, tape, mechanical photocopying, recording or otherwise without the written permission of the Publisher.

We have partnered with Copyright Clearance Center to make it easy for you to obtain permissions to reuse content from this publication. Simply navigate to this publication's page on Nova's website and locate the "Get Permission" button below the title description. This button is linked directly to the title's permission page on copyright.com. Alternatively, you can visit copyright.com and search by title, ISBN, or ISSN.

For further questions about using the service on copyright.com, please contact:
Copyright Clearance Center
Phone: +1-(978) 750-8400   Fax: +1-(978) 750-4470   E-mail: info@copyright.com.

NOTICE TO THE READER

The Publisher has taken reasonable care in the preparation of this book, but makes no expressed or implied warranty of any kind and assumes no responsibility for any errors or omissions. No liability is assumed for incidental or consequential damages in connection with or arising out of information contained in this book. The Publisher shall not be liable for any special, consequential, or exemplary damages resulting, in whole or in part, from the readers' use of, or reliance upon, this material. Any parts of this book based on government reports are so indicated and copyright is claimed for those parts to the extent applicable to compilations of such works.

Independent verification should be sought for any data, advice or recommendations contained in this book. In addition, no responsibility is assumed by the Publisher for any injury and/or damage to persons or property arising from any methods, products, instructions, ideas or otherwise contained in this publication.

This publication is designed to provide accurate and authoritative information with regard to the subject matter covered herein. It is sold with the clear understanding that the Publisher is not engaged in rendering legal or any other professional services. If legal or any other expert assistance is required, the services of a competent person should be sought. FROM A DECLARATION OF PARTICIPANTS JOINTLY ADOPTED BY A COMMITTEE OF THE AMERICAN BAR ASSOCIATION AND A COMMITTEE OF PUBLISHERS.

Additional color graphics may be available in the e-book version of this book.

Library of Congress Cataloging-in-Publication Data

ISBN: 978-1-53616-020-8
Library of Congress Control Number: 2019946603

Published by Nova Science Publishers, Inc. † New York

CONTENTS

Preface		**vii**
Chapter 1	Analysis of Preferential Solvation and Hydration of Binary and Ternary Mixtures: Methodology *Vladimir A. Sirotkin*	**1**
Chapter 2	Preferential Solvation of α-Chymotrypsin in Water-Monohydric Alcohol Mixtures: Effect of Chain Length *Vladimir A. Sirotkin and Alexandra A. Kuchierskaya*	**25**
Chapter 3	Preferential Hydration of α-Chymotrypsin in Acetonitrile: Comparison with FTIR Spectroscopy *Vladimir A. Sirotkin*	**57**
Chapter 4	Preferential Solvation of Lysozyme in Water-Dimethyl Sulfoxide Mixtures: Gibbs Energies of Water and Organic Solvent *Vladimir A. Sirotkin*	**91**

Chapter 5	Preferential Solvation and Hydration of Lysozyme in Ethylene Glycol and Ethanol: Effect of Hydroxyl Group *Vladimir A. Sirotkin*	**123**
Author's Contact Information		**149**
Index		**151**
Related Nova Publications		**159**

PREFACE

This book describes the basic principles of a novel methodology to investigate the preferential hydration and solvation of proteins in ternary protein-water-organic solvent systems. Protein-water interactions are well-known to play a critical role in determining the function, structure, and stability of protein macromolecules. Elucidation of the processes occurring upon protein hydration in the presence of third component (organic solvents, salts, urea) is essential in a wide range of biophysical, biomedical, and biotechnological applications. In particular, there are many advantages in employing water-poor organic solvents, including the suppression of undesirable side reactions caused by water, the biocatalysis of reversed hydrolytic reactions (transesterification), or increased thermostability. Distinct intermediate protein states induced by organic solvents may be responsible for numerous neurodegenerative diseases (Alzheimer's disease, Parkinson's disease, and Huntington's disease). However, the manner in which organic solvents increase/decrease the thermal stability, induce/reduce the extent of denaturation, and stabilize/destabilize the partially folded conformations of proteins (amyloid fibrils and molten globules) is an intricate function of water content in organic liquids.

Preferential hydration/solvation is an effective method for revealing the mechanism of the protein stabilization or denaturation. When a protein interacts with a binary water-organic solvent mixture, the three components do not equally mix. Water or organic solvent molecules exist preferentially in the protein's solvation shell. This difference between the solvation shell and bulk solvent in the solvent components has been termed preferential solvation. Preferential solvation is a thermodynamic quantity that describes the protein surface occupancy by the water and cosolvent molecules. This is associated with the actual numbers of water/cosolvent molecules that are in contact with the protein's surface. It was also found that the protein destabilization is directly associated with the preferential binding of the denaturant molecules to specific protein groups.

The aim of our study is to monitor the preferential solvation and preferential hydration of the protein macromolecules at low, intermediate, and high water content in organic solvents at 25 °C. Our approach is based on the simultaneous measurements of the absolute values of the water and organic solvent sorption. The preferential solvation/hydration parameters were calculated using the water and organic solvent sorption values. The preferential solvation/hydration parameters were compared with the corresponding changes in the protein structure that transpire regarding the interaction of the protein with organic solvent and water molecules. The effect of organic solvent on the protein structure was investigated by FTIR (Fourier Transform Infrared) spectroscopy.

Advantages of our approach:

(i) The preferential interaction parameters can be determined in the entire range of water content in organic liquids.
(ii) Our approach facilitates the individual evaluation of the Gibbs energies of water, protein, and organic solvent.

Chapter 1

ANALYSIS OF PREFERENTIAL SOLVATION AND HYDRATION OF BINARY AND TERNARY MIXTURES: METHODOLOGY

ABSTRACT

This chapter describes the basic principles of a novel methodology to investigate preferential interactions in binary and ternary mixtures. Our methodology for binary mixtures is based on an analysis of the densities of water-organic solvent and water-protein systems. Our approach for ternary water-protein-organic solvent mixtures is based on a simultaneous analysis of absolute values of the water and organic solvent sorption. Advantages of our approach: (i) The preferential interaction parameters can be determined in the entire range of water content in organic liquids. (ii) Our approach facilitates the individual evaluation of the Gibbs energies of water and organic solvent.

Keywords: methodology, preferential solvation and hydration, water, organic solvent, binary mixtures, thermodynamic functions

1. INTRODUCTION

The aim of this investigation is to present a novel methodology for studying preferential solvation and preferential hydration of the protein macromolecules at low, intermediate, and high water content in organic liquids at 25°C.

Owing to the ability to vary the size, polarity, and strength of hydrogen bonds, organic solvents, including alcohols, are used in biophysical chemistry and biotechnology to selectively modulate the properties of a protein system. As an example of an innovation-promising scientific area, enzymatic catalysis in nonaqueous media (including organic solvents, ionic liquids, and supercritical fluids) has been intensively developed, recently [1-13]. There are several advantages in employing nonaqueous media for biocatalysis, including the high solubility of hydrophobic reagents, the synthesis of useful chemicals, the suppression of undesirable side reactions caused by water and enhanced thermostability. High selectivity (perhaps the most attractive feature of enzymes in organic liquids) can be markedly affected, and sometimes even inverted, by the solvent. Notably, the protein activity and structure in organic solvents depend in a complicated way on the water content.

Preferential solvation is an effective way for revealing the mechanism of protein denaturation or stabilization [14-29]. When a protein interacts with a binary aqueous-organic mixture, the three components do not equally mix. Water or organic solvent molecules exist preferentially in the protein's solvation shell. This difference between the bulk solvent and solvation shell in the solvent components has been termed preferential solvation. Preferential solvation is a thermodynamic quantity that describes the protein molecule surface occupancy by the cosolvent and water molecules.

Our approach for ternary water-protein-organic solvent mixtures is based on a simultaneous analysis of absolute values of the water and organic solvent sorption. One of the most important advantages of our approach is the facilitation of individual evaluation of the Gibbs energies of water, organic solvent, and protein in the entire range of water content.

Our methodology for binary mixtures is based on the analysis of the densities of water-organic solvent and water-protein systems. It facilitates the individual evaluation of the organic solvent and water partial quantities in the entire range of water content.

2. METHODOLOGY. THERMODYNAMIC FUNCTIONS OF BINARY MIXTURES

2.1. Partial Quantities

A partial property is a thermodynamic quantity which indicates how an extensive property (volume, enthalpy, Gibbs energy) of a mixture varies with changes in the composition of the mixture at constant temperature and pressure [30-33]. Essentially it is the partial derivative of the extensive property with respect to the amount (number of moles) of the component of interest (F_i). Every extensive property of a mixture has a corresponding partial property (Eq. 1):

$$F_i = \left(\frac{\partial F}{\partial n_i}\right)_{P,T,n_j} \tag{1}$$

For binary mixtures. the partial quantities of components 1 and 2 are calculated using Eqs. 2 and 3:

$$F_1^M = F^M - x_2\left(\frac{dF^M}{dx_2}\right)_{T,P} \tag{2}$$

$$F_2^M = F^M - x_1\left(\frac{dF^M}{dx_1}\right)_{T,P} \tag{3}$$

Where x_1 and x_2 are the mol fractions of components 1 and 2, respectively ($x_1 + x_2 = 1$).

The partial volumes of water (component 1) and component 2 are calculated using Eqs. 4 and 5:

$$V_1^M = V^M - x_2 \left(\frac{dV^M}{dx_2}\right)_{T,P} \tag{4}$$

$$V_2^M = V^M - x_1 \left(\frac{dV^M}{dx_1}\right)_{T,P} \tag{5}$$

The partial Gibbs energies (chemical potentials) of binary mixtures are calculated using Eqs. 6 and 7:

$$\mu_1 = \mu_1^0 + RT \ln x_1 \tag{6}$$

$$\mu_2 = \mu_2^0 + RT \ln x_2 \tag{7}$$

2.2. Excess Partial Quantities

The thermodynamic properties (volume V, enthalpy H, entropy S, and Gibbs free energy G) of a real binary system can be expressed in terms of the excess functions, F^E. They are the difference between the thermodynamic function of mixing in a real system and the value corresponding to an ideal system at the same temperature, pressure, and composition. For an ideal system, all excess functions are zero. Deviations of the excess functions from zero indicate the extent to which the studied binary system is non-ideal due to strong specific interactions between components (i.e., hydrogen bonding and charge-charge interactions). More comprehensive reviews of excess functions are given in Refs. 30-33.

The F^E values can be calculated using Eq. 8:

$$F^E = F^M - F^M_{id} \tag{8}$$

where F^M is the thermodynamic function of mixing for a real system; F^M_{id} is the thermodynamic function for an ideal mixture.

For ideal binary mixtures, the F^M_{id} values can be calculated using Eq. 9:

$$F^M_{id} = x_1 F^o_1 + x_2 F^o_2 \tag{9}$$

where F^o_1 and F^o_2 are the thermodynamic function values for pure water and pure second component, and x_1 and x_2 are the mol fractions of water and second component, respectively ($x_1 + x_2 = 1$).

For real binary mixtures, the F^E values are composed of two components:

$$F^E = x_1 F^E_1 + x_2 F^E_2 \tag{10}$$

where F^E_1 and F^E_2 are the excess partial quantities of components 1 and 2.

The excess partial quantities of components 1 and 2 are calculated using Eqs. 11 and 12:

$$F^E_1 = F^E - x_2 \left(\frac{dF^E}{dx_2}\right)_{T,P} \tag{11}$$

$$F^E_1 = F^E - x_2 \left(\frac{dF^E}{dx_2}\right)_{T,P} \tag{12}$$

The excess partial volumes of water and second component are calculated using Eqs. 13 and 14:

$$V^E_1 = V^E - x_2 \left(\frac{dV^E}{dx_2}\right)_{T,P} \tag{13}$$

$$V_2^E = V^E - x_1 \left(\frac{dV^E}{dx_1}\right)_{T,P} \tag{14}$$

The excess partial Gibbs energies (excess chemical potentials) of water and component 2 are calculated using Eqs. 15 and 16:

$$\mu_1^E = RT\ln\gamma_1 \tag{15}$$

$$\mu_2^E = RT\ln\gamma_2 \tag{16}$$

Where γ_1 are γ_2 are the thermodynamic activity coefficients.

3. ANALYSIS OF PREFERENTIAL SOLVATION IN BINARY MIXTURES

3.1. Methodology

The partial quantities of the components of a mixture vary with the composition of the mixture, because the environment of the molecules in the mixture changes with the composition. It is the change of the molecular environment (and the consequent alteration of *the preferential interactions* between molecules) that results in the thermodynamic properties of a mixture changing as its composition is altered.

Preferential solvation (hydration) in binary mixtures was defined as follows. The absolute values of the solvation (hydration) excess ($\Gamma_{A(B)}^k$) were estimated using Eqs. 17-21:

$$\Gamma_{1(2)}^k = -\left(\frac{d\ln[\gamma_1(pref.hydration)]}{d\ln(a_1)}\right)_{T,P,a_2} \tag{17}$$

$$\Gamma_{2(1)}^k = -\left(\frac{d\ln[\gamma_2(pref.solvation)]}{d\ln(a_2)}\right)_{T,P,a_1} \tag{18}$$

where $\Gamma_{A(B)}^k$ is the solvation excess of the A particles over the B component in the surrounding of k particles [33]; ($\gamma_1(binary\ mixture)$ and $\gamma_2(binary\ mixture)$ are the activity coefficients in the unperturbed binary mixture; $\gamma_1(solvation\ layer)$ and $\gamma_2(solvation\ layer)$ are the activity coefficients in the solvation layer; a_1 and a_2 are the thermodynamic activities. Water activity (a_1) and organic solvent activity (a_2) were taken from the published data [34] based on the vapor-liquid equilibrium.

The physical representation of solvation excess is determined by Eq. 19:

$$\Gamma_{A(B)}^k = \frac{N_A}{V}\int_0^{r_c}(\rho_{kA} - \rho_{kB}) * 4\pi r^2 dr \tag{19}$$

where r_c is the molecular correlation radius; N_A is the number of "A" particles in the volume V, and ρ_{kA} (ρ_{kB}) is the radial distribution function.

The solvation excess ($\Gamma_{A(B)}^k$) is the average number of the A particles by which the solvation shell of the k particles is richer as compared to the B molecules. If in the solvent surrounding a k particle the ratio of the A and B particle numbers is equal to their ratio in the unperturbed bulk solution, then $\Gamma_{A(B)}^k = 0$. No preferential solvation (hydration) occurs in this case.

The solvation excess ($\Gamma_{A(B)}^k$) is analogous to the Gibbs sorption.

The hydration excess values ($\Gamma_{1(2)}^1$) in various water-organic mixtures are given in Figures 1-3. There are three distinct mixing schemes.

One composition regime was observed in water-glycerol and water-formamide mixtures (Figure 1). As concluded from Figure 1, the hydration excess values are positive in the entire range of water content in glycerol (a hydrophilic structure maker, protein stabilizer). On the other hand, the hydration excess values are negative in the entire range of water content in formamide (a hydrophilic structure breaker, protein destabilizer).

Two composition regimes were observed in water-ethylene glycol (Figure 2) and water-DMSO (Figure 3) mixtures. As shown in Figures 2 and 3, the hydration excess values are positive at high water content. This means that the solvation shell of water molecules is richer by the water molecules as compared to the unperturbed bulk solvent.

At low water content, however, the $\Gamma^1_{1(2)}$ values are negative. This fact indicates that the solvation shell of water molecules is richer by the DMSO (ethylene glycol) molecules as compared to the unperturbed bulk solvent.

Three composition regions were detected in water-acetonitrile mixtures (Figure 3). As shown in Figure 3, the hydration excess values are positive at the highest water content ($w_1 = 0.85 - 1.0$).

At intermediate water content, however, the $\Gamma^1_{1(2)}$ values are negative. This means that the solvation shell of water molecules is richer by the acetonitrile molecules as compared to the unperturbed bulk solvent.

At the lowest water content, the $\Gamma^1_{1(2)}$ values are positive. This means that the solvation shell of water molecules is richer by the water molecules as compared to the unperturbed bulk solvent.

Thermodynamic functions of the preferential interactions in binary water-organic mixtures were determined using Eqs. 20 and 21. These equations were utilized to determine the Gibbs energies of the transfer of water [$\Delta G^{pref}_1 = RT\ln[\gamma_1(pref.hydration)]$] and component 2 [$\Delta G^{pref}_2 = RT\ln[\gamma_2(pref.solvation)]$] from the unperturbed binary mixture to the solvation layer of water or component 2:

$$\Delta G^{pref}_1 = \mu^E_1(solvation\ layer) - \mu^E_1(binary\ mixture) \qquad (20)$$

$$\Delta G_2^{pref} = \mu_2^E(solvation\ layer) - \mu_2^E(binary\ mixture) \tag{21}$$

The $\mu_1^E(binary\ mixture)$ and $\mu_2^E(binary\ mixture)$ values can be calculated using Eqs. 22 and 23:

$$\mu_1^E(binary\ mixture) = RTln\gamma_1(binary\ mixture) \tag{22}$$

$$\mu_2^E(binary\ mixture) = RTln\gamma_2(binary\ mixture) \tag{23}$$

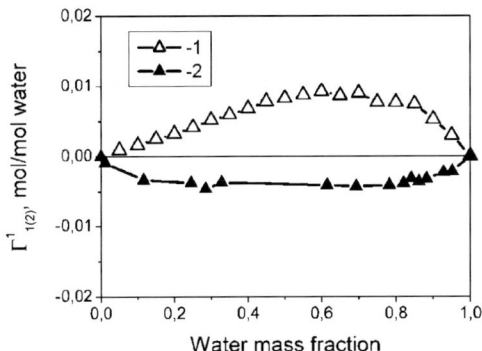

Figure 1. The hydration excess for water-glycerol (1) and water-formamide (2) mixtures at 25°C.

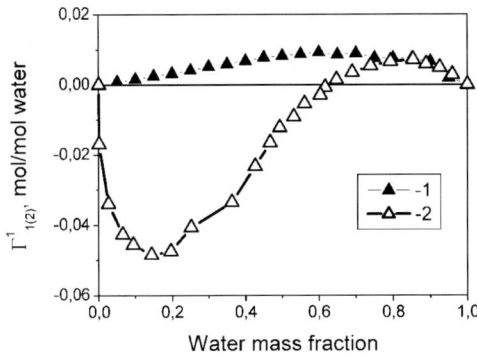

Figure 2. The hydration excess for water-glycerol (1) and water-ethylene glycol (2) mixtures at 25°C.

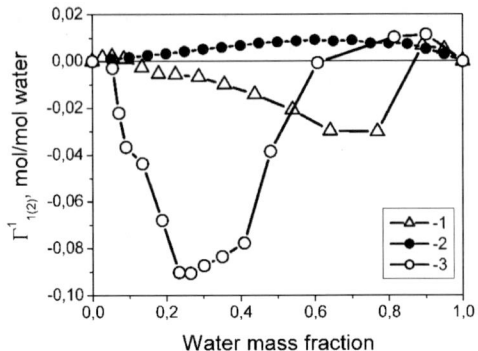

Figure 3. The hydration excess for water-acetonitrile (1), water-glycerol (2), water-DMSO (3) mixtures at 25°C.

Water activity coefficients ($\gamma_1(binary\ mixture)$, the mass fraction scale; the reference state is pure water) in aqueous-organic mixtures were estimated using Eq. 24:

$$\gamma_1(binary\ mixture) = \frac{a_1}{w_1} \tag{24}$$

Organic solvent activity coefficients ($\gamma_2(binary\ mixture)$, the mass fraction scale; the reference state is pure organic solvent) in aqueous-organic mixtures were calculated using Eq. 25:

$$\gamma_2(binary\ mixture) = \frac{a_2}{w_2} \tag{25}$$

The $\mu_1^E(solvation\ layer)$ and $\mu_2^E(solvation\ layer)$ values were calculated using Eqs. 26 and 27:

$$\mu_1^E(solvation\ layer) = RT ln\gamma_1(solvation\ layer) \tag{26}$$

$$\mu_2^E(solvation\ layer) = RT ln\gamma_2(solvation\ layer) \tag{27}$$

Water activity coefficients ($\gamma_1(solvation\ layer)$, the mass fraction scale) in the solvation layer were calculated using Eq. 28:

$$\gamma_1(solvation\ layer) = \frac{a_1}{Z_1^M} \qquad (28)$$

where Z_1^M is the mass fraction of water in the solvation layer; ρ_1^0 – the density of pure water:

$$Z_1^M = \frac{\rho_1}{\rho_1^0} w_1.$$

Organic solvent activity coefficients ($\gamma_2(solvation\ layer)$, the mass fraction scale) in the solvation layer were estimated using Eq. 29:

$$\gamma_2(solvation\ layer) = \frac{a_2}{Z_2^M} \qquad (29)$$

where Z_2^M is the organic solvent mass fraction in the solvation layer; ρ_2^0 is the density of pure organic solvent:

$$Z_2^M = \frac{\rho_2}{\rho_2^0} w_2.$$

Eqs. 30-32 were used to calculate the partial densities of water (component 1) and the second component (organic solvent or protein) in the binary mixtures:

$$\rho^M = w_1 \rho_1 + w_2 \rho_2 \qquad (30)$$

$$\rho_1 = \rho^M - w_2 \left(\frac{d\rho^M}{dw_2}\right)_{T,P} \qquad (31)$$

$$\rho_2 = \rho^M - w_1 \left(\frac{d\rho^M}{dw_1}\right)_{T,P} \tag{32}$$

where ρ^M (g/cm³) is the density of binary water-organic or water-protein mixture; ρ_1 and ρ_2 are the partial densities of water and component 2, respectively. w_1 is the water mass fraction in binary mixtures; and w_2 is the mass fraction of component 2 in binary mixtures ($w_1 + w_2 = 1.0$).

The Gibbs energies of the transfer of water (ΔG_1^{pref}) and organic solvent (ΔG_2^{pref}) from the water-dimethyl sulfoxide (DMSO) mixtures to the solvation layer of water and DMSO are presented in Figure 2.

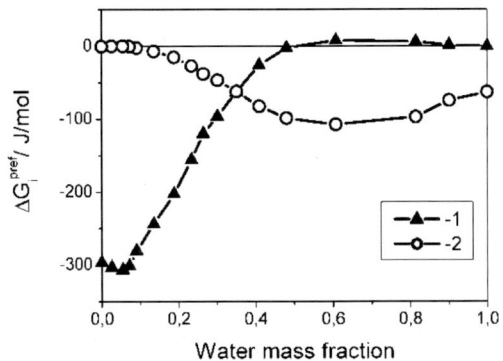

Figure 4. The Gibbs energies of the transfer of water (ΔG_1^{pref}) and organic solvent (ΔG_2^{pref}) from the water-dimethyl sulfoxide (DMSO) mixtures to the solvation layer of water and DMSO at 25°C.

3.2. Densitometry

Density measurements of pure liquids and mixtures were performed at atmospheric pressure and 25°C by means of vibrating-tube densimeter (Anton Paar, Austria, DMA 5000M, precision ± 1 × 10⁻⁶ g cm⁻³). Before each series of measurements the densimeter was calibrated with distilled water and air. All water-organic mixtures were prepared gravimetrically using a Precisa balance (Swiss) with a precision of 0.00001 g.

4. ANALYSIS OF TERNARY WATER-PROTEIN-ORGANIC SOLVENT MIXTURES

4.1. Methodology

The protein solvation layer is composed of two parts: (i) non-ideal (due to preferential interactions) and (ii) ideal. The non-ideal effect of the solvation layer on protein properties (organic solvent/water sorption; residual enzyme activity) can be defined in terms of excess functions, F^E [31-33], which refers to the difference between the observed function of mixing, F^M, and the function for an ideal binary mixture, F_{id}^M.

The deviations of the excess functions from zero describe the degree to which the protein solvation layer differs from the pure binary water-organic mixture as a consequence of preferential interactions between water (component 1), protein (component 2), and organic solvent (component 3).

Eq. 33 is used to calculate the F^E values:

$$F^E = F^M - F_{id}^M \tag{33}$$

The $F_{id,i}^M$ values were calculated using Eq. 34:

$$F_{id,i}^M = F_i^M(w_i = 0) + w_i[F_i^M(w_i = 1.0) - F_i^M(w_i = 0)] \tag{34}$$

where $F_i^M(w_i = 1.0)$ is the observed mixing function of protein at $w_i = 1.0$; $F_i^M(w_i = 0)$ is the observed mixing function of protein at $w_i = 0$;

w_1 is the water mass fraction in the binary water-organic mixtures; and w_3 is the organic solvent mass fraction in the binary water-organic mixtures ($w_1 + w_3 = 1.0$).

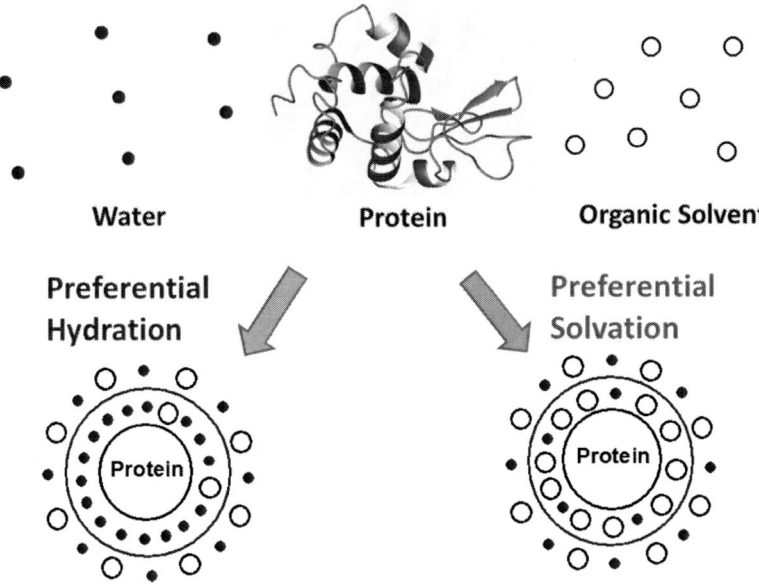

Figure 5. Scheme describing preferential interactions between water, protein, and organic solvent.

The situation in which no preferential interactions exist between water, protein, and organic solvent is described by the $F_{id,i}^M$ values. In this circumstance, the water mass fraction in the ideal part of the solvation layer is the same as in the pure water-organic mixture.

The Z_1^M (water mass fraction in the solvation layer) and Z_3^M (organic solvent mass fraction in the solvation layer) values as a function of water mass fraction in organic solvent were calculated using Eqs. 35 and 36, respectively:

$$Z_1^M = \frac{A_1}{A_1 + A_3} \qquad (35)$$

$$Z_3^M = \frac{A_3}{A_1 + A_3} \tag{36}$$

The simultaneous action of organic solevent and water was determined by the Z_1^E and Z_3^E values. These excess sorption functions can be estimated using Eqs. 37 and 38:

$$Z_1^E = Z_1^M - Z_{id,1}^M \tag{37}$$

$$Z_3^E = Z_3^M - Z_{id,3}^M \tag{38}$$

where Z_1^M is the mass fraction of water in the solvation layer for the real water-organic mixture; and $Z_{id,1}^M$ is the mass fraction of water in the solvation layer for the ideal water-organic mixture. The $Z_{id,1}^M$ values were calculated using Eq. 39:

$$Z_{id,1}^M = Z_1^M(w_1 = 0) + w_1[Z_1^M(w_1 = 1.0) - Z_1^M(w_1 = 0)] \tag{39}$$

where $Z_1^M(w_1 = 1.0)$ is the water mass fraction in the solvation layer of protein at w_1=1.0; $Z_1^M(w_1 = 0)$ is the water mass fraction in the solvation shell at w_1=0; and w_1 is the mass fraction of water in organic solvent.

In addition, Z_3^M is the organic solvent mass fraction in the solvation layer for the real water-organic mixture; and $Z_{id,3}^M$ is the organic solvent mass fraction for the ideal water-organic mixture. Eq. 40 is used to calculate the $Z_{id,3}^M$ values:

$$Z_{id,3}^M = Z_3^M(w_3 = 0) + w_3[Z_3^M(w_3 = 1.0) - Z_3^M(w_3 = 0)] \tag{40}$$

where $Z_3^M(w_3 = 0)$ is the organic mass fraction in the solvation shell of protein at w_3=1.0; $Z_3^M(w_3 = 0)$ is the organic solvent mass fraction in the solvation shell of protein at $w_3=0$; and w_3 is the organic solvent mass fraction in the binary water-organic mixture.

The degree to which the solvation layer differs from the pure binary water-organic mixture due to preferential interactions between water (component 1), protein (component 2), and organic solvent (component 3) can be characterized by the preferential interaction parameters (Eqs. 41 and 42). Eq. 41 [14-16] was utilized to determine the preferential binding of protein in water-organic mixtures:

$$(\partial g_3/\partial g_2)_{T,\mu 1,\mu 3} = A_3 - \frac{w_3}{w_1} A_1 \qquad (41)$$

where A_1 is the protein hydration, expressed as gram water per gram protein; A_3 is the binding of organic solvent, expressed as gram organic solvent per gram protein; w_1 is the water mass fraction in water-organic mixtures; and w_3 is the mass fraction of organic solvent in water-organic mixtures ($w_1 + w_3 = 1.0$).

Eq. 12 was used to estimate the preferential hydration:

$$(\partial g_1/\partial g_2)_{T,\mu 1,\mu 3} = -\left(\frac{w_1}{w_3}\right)(\partial g_3/\partial g_2)_{T,\mu 1,\mu 3} \qquad (42)$$

Eqs. 43 and 44 were utilized to determine the Gibbs energies of the transfer of water (ΔG_1^{pref}) and acetonitrile (ΔG_3^{pref}) from water-organic mixtures to the solvation layer of protein:

$$\Delta G_1^{pref} = \mu_1^E(solvation\ layer) - \mu_1^E(binary\ mixture) \qquad (43)$$

$$\Delta G_3^{pref} = \mu_3^E(solvation\ layer) - \mu_3^E(binary\ mixture) \qquad (44)$$

The $\mu_1^E(binary\ mixture)$ and $\mu_3^E(binary\ mixture)$ values can be calculated using Eqs. 45 and 46:

$$\mu_1^E(binary\ mixture) = RTln\gamma_1(binary\ mixture) \tag{45}$$

$$\mu_3^E(binary\ mixture) = RTln\gamma_3(binary\ mixture) \tag{46}$$

Water activity coefficients ($\gamma_1(binary\ mixture)$, the mass fraction scale; the reference state is pure water) in aqueous-organic mixtures were estimated using Eq. 47:

$$\gamma_1(binary\ mixture) = \frac{a_1}{w_1} \tag{47}$$

Organic solvent activity coefficients ($\gamma_3(binary\ mixture)$, the mass fraction scale; the reference state is pure alcohol) in aqueous-organic mixtures were calculated using Eq. 18:

$$\gamma_3(binary\ mixture) = \frac{a_3}{w_3} \tag{48}$$

Water activity (a_1) and organic solvent activity (a_3) were taken from the published data [34] based on the vapor-liquid equilibrium.

The $\mu_1^E(solvation\ layer)$ and $\mu_3^E(solvation\ layer)$ values were calculated using Eqs. 49 and 50:

$$\mu_1^E(solvation\ layer) = RTln\gamma_1(solvation\ layer) \tag{49}$$

$$\mu_3^E(solvation\ layer) = RTln\gamma_3(solvation\ layer) \tag{50}$$

Water activity coefficients ($\gamma_1(solvation\ layer)$, the mass fraction scale) in the solvation layer were calculated using Eq. 51:

$$\gamma_1(solvation\ layer) = \frac{a_1}{Z_1^M} \tag{51}$$

where Z_1^M is the mass fraction of water in the solvation layer;

$$Z_1^M = \frac{A_1}{A_1+A_3}.$$

Organic solvent activity coefficients ($\gamma_3(solvation\ layer)$, the mass fraction scale) in the solvation layer were estimated using Eq. 52:

$$\gamma_3(solvation\ layer) = \frac{a_3}{Z_3^M} \qquad (52)$$

where Z_3^M is the organic solvent mass fraction in the solvation layer;

$$Z_3^M = \frac{A_3}{A_1+A_3}.$$

ΔG_2^{pref} values described protein stabilization/destabilization due to the preferential interactions. The Gibbs-Duhem equation for ternary systems (Eq. 53) was utilized to calculate the ΔG_2^{pref} values:

$$\Delta G_2^{pref} = \frac{m_1 \Delta G_1^{pref} + m_3 \Delta G_3^{pref}}{m_2} \qquad (53)$$

$$\Delta G_2^{pref} = \mu_2^E(protein\ in\ water - organic\ mixtures) - \mu_2^E(pure\ protein) \qquad (54)$$

where μ_1^E, μ_2^E, and μ_3^E are the excess chemical potentials of water, protein, and organic, respectively; and m_1, m_2, and m_3 are the masses of water, protein, and organic solvent, respectively.

4.2. Sorption Measurements

The protein samples were prepared according to the recommended guidelines [35-38]. The initially dried protein samples were presented to water-organic solvent vapor mixtures. The water-organic vapor mixture was consecutively flowed through a thermostated saturator filled with the water-organic 1 mixture, and a cell containing the protein sample. Protein samples (5-10 mg) each were flushed by water – organic vapor mixtures, until no further mass changes were detected as described previously. The sorption equilibrium was reached after 6 h at 25°C. The schematic representation of the experimental setup is given in Ref. 35. An external ethylene glycol thermostat (RC 6, Lauda, Germany) was utilized to determine the temperature with a precision of 0.1°C. The water activity (a_1) in the vapor phase was adjusted by altering the water content in the liquid water-organic mixture.

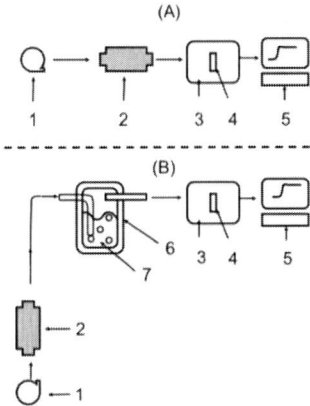

Figure 6. Schematic presentation of the experimental set-up of the sorption measurements. The components of the experimental set-up: **(A)** 1 – air pump; 2 - thermostated glass tube with P_2O_5; 3 – microthermoanalyzer "Setaram" MGDTD-17S; 4 – thermostated cell; 5 – Karl Fischer titrator. **(B)** *1* – air pump; 2 - thermostated glass tube with P_2O_5; 3 – microthermoanalyzer "Setaram" MGDTD-17S; 4 – thermostated cell; 5 – Karl Fischer titrator; 6 – thermostated saturator; 7 – water - organic mixture.

Measurements of the water sorption (A_1) were conducted by Karl Fischer titration with a Metrohm 831 KF coulometer. Organic solvent content of protein (A_3) was calculated as a difference between the total sorption uptake ($A_1 + A_3$) and water content (A_1). The total sorption uptake ($A_1 + A_3$) was measured by microthermoanalyzer "Setaram" MGDTD-17S.

REFERENCES

[1] Klibanov, A. M. (2001). Improving enzymes by using them in organic solvents, *Nature*, *409*, 241-246.

[2] Carrea, G. & Riva, S. (2000). Properties and synthetic applications of enzymes in organic solvents, *Angew. Chem. Int. Ed.*, *39*, 2226-2254.

[3] Halling, P. J. (2004). What can we learn by studying enzymes in nonaqueous media? *Phil. Trans. R. Soc. Lond. B Biol. Sci.*, *359*, 1287– 1297.

[4] Micaelo, N. M. & Soares C. M. (2007). Modeling hydration mechanisms of enzymes in nonpolar and polar organic solvents, *FEBS J.*, *274*, 2424–2436.

[5] Clark, D. S. (2004). Characteristics of nearly dry enzymes in organic solvents: implications for biocatalysis in the absence of water, *Phil. Trans. R. Soc. Lond. B Biol. Sci.*, *359*, 1299–1307.

[6] Serdakowski, A. L. & Dordick, J. S. (2007). Enzyme activation for organic solvents made easy, *Trends Biotechnol.*, *26*, 54-48.

[7] Rariy, R. & Klibanov, A. M. (1997) Correct protein folding in glycerol, *Proc. Natl. Acad. Sci. U.S.A.*, *94*, 13520-13523.

[8] Sirotkin, V. A., Hüttl, R. & Wolf, G., (2008). Enzyme-catalysed hydrolysis of L-amino acid esters in a low water organic solvent studied by isothermal calorimetry, *J. Therm. Anal. Calorim*, *93*, 515-520.

[9] Sirotkin, V. A. & Faizullin, D. A. (2004). Interaction enthalpies of solid human serum albumin with water-dioxane mixtures: comparison with water and organic solvent vapor sorption, *Thermochim. Acta*, *415*, 127-133.

[10] Partridge, J., Hutcheon, G. A., Moore, B. D. & Halling P. J. (1996). Exploiting hydration hysteresis for high activity of cross-linked subtilisin crystals in acetonitrile, *J. Amer. Chem. Soc.*, *118*, 12873-12877.

[11] Kijima, T., Yamamoto, S. & Kise, H. (1996). Study of tryptophan fluorescence and catalytic activity of α-chymotrypsin in aqueous organic media, *Enz. Microb. Technol.*, *18*, 2-6.
[12] Khmelnitsky, Yu., Mozhaev, V. V., Belova, A. B., Sergeeva, M. V. & Martinek, K. (1991). Denaturation capacity: a new quantitative criterion for selection of organic solvents as reaction media in biocatalysis, *Eur. J. Biochem.*, *198*, 31-41.
[13] Simon, L. M., Kotorman, M., Garab, G. & Laczko, I. (2001). Structure and activity of δ-chymotrypsin and trypsin in aqueous organic media, *Biochem. Biophys. Res. Comm.*, *280*, 1367-1371.
[14] Timasheff, S. N. (2002). Protein-solvent preferential interactions, protein hydration, and the modulation of biochemical reactions by solvent components. *Proc. Natl. Acad. Sci. USA.*, *99*, 9721-9726.
[15] Arakawa, T., Kita, Y. & Timasheff, S. N. (2007). Protein precipitation and denaturation by dimethyl sulfoxide. *Biophys. Chem.*, *131*, 62-70.
[16] Gekko, K., Ohmae, E., Kameyama, K. & Takagi, T. (1998). Acetonitrile-protein interactions: Amino acid solubility and preferential solvation. *Biochim. Biophys. Acta*, *1387*, 195-205.
[17] Kovrigin, E. L. & Potekhin, S. A. (2000). On stabilizing action of protein denaturants: Acetonitrile effect on stability of lysozyme in aqueous solutions. *Biophys. Chem.*, *83*, 45-59.
[18] Shimizu, S. & Matubayasi, N. (2014). Preferential solvation: Dividing surface vs excess numbers. *J. Phys. Chem. B.*, *118*, 3922-3930.
[19] Auton, M., Bolen, D. W. & Rosgen, J. (2008). Structural thermodynamics of protein preferential solvation: Osmolyte solvation of proteins, amino acids, and peptides. *Proteins*, *73*, 802–813.
[20] Casassa, E. F. & Eisenberg, H. (1964). Thermodynamic analysis of multicomponent solutions. *Adv. Protein Chem.*, *19*, 287–395.

[21] Tanford, C. (1969). Extension of theory of linked functions to incorporate effects of protein hydration. *J. Mol. Biol.*, *39*, 539−544.
[22] Schellman, J. A. (1987). Selective binding and solvent denaturation. *Biopolymers*, *26*, 549−559.
[23] Arakawa, T., Bhat, R. & Timasheff, S. N. (1990). Why preferential hydration does not always stabilize the native structure of globular proteins. *Biochem.*, *29*, 1924-1931.
[24] Smith, P. E. (2006). Equilibrium dialysis data and the relationships between preferential interaction parameters in biological systems in terms of Kirkwood-Buff integrals. *J. Phys. Chem. B*, *110*, 2862− 2868.
[25] Parsegian, V. A., Rand, R. P. & Rau, D. C. (2000). Osmotic stress, crowding, preferential hydration, and binding: A comparison of perspectives. *Proc. Natl. Acad. Sci. U.S.A.*, *97*, 3987−3992.
[26] Kirby Hade, E. P. & Tanford, C. (1967). Isopiestic compositions as a measure of preferential interactions of macromolecules in two-component solvents. Application to proteins in concentrated aqueous cesium chloride and guanidine hydrochloride. *J. Amer. Chem. Soc.*, *89*, 5034-5040.
[27] Kamiyama, T., Liu, H. L. & Kimura, T. (2009). Preferential solvation of lysozyme by dimethyl sulfoxide in binary solutions of water and dimethyl sulfoxide. *J. Therm. Anal. Cal.*, *95*, 353-359.
[28] Reisler, E., Haik, Y. & Eisenberg, H. (1977). Bovine serum albumin in aqueous guanidine hydrochloride solutions. Preferential and absolute interactions and comparison with other systems. *Biochem.*, *16*, 197-203.
[29] Izumi, T., Yoshimura, Y. & Inoue, H. (1980). Solvation of lysozyme in water/dioxane mixtures studied in the frozen state by NMR spectroscopy. *Arch. Biophys. Biochem.*, *200*, 444-451.
[30] Atkins, P. W. (2006). *Physical Chemistry*. 8th ed. Oxford: Oxford University Press.

[31] Prausnitz, J. M. (1969). *Molecular Thermodynamics of Fluid-Phase Equilibria*. N.J.: Prentice-Hall, Inc., Engelwood Cliffs.
[32] Belousov, V. P. & Panov, M. Y. (1994). *Thermodynamic properties of aqueous solutions of organic substances*. Boca Raton, Fla.: CRC Press.
[33] Pendin, A. A. (1989). Preferential solvation and thermodynamical properties of nonelectrolites solutions. *Russ. J. Phys. Chem.*, *63*, 1793-1798.
[34] Bell, G., Janssen, A. E. M. & Halling, P. (1996). Water activity fails to predict critical hydration level for enzyme activity in polar organic solvents: Interconversion of water concentrations and activities. *Enzym. Microb. Technol.*, *20*, 471-476.
[35] Sirotkin, V. A. & Kuchierskaya, A. A. (2017). Preferential solvation/hydration of α-chymotrypsin in water-acetonitrile mixtures. *J. Phys. Chem. B.*, *121*, 4422-4430.
[36] Sirotkin, V. A. & Kuchierskaya, A. A. (2017). Lysozyme in water-acetonitrile mixtures: Preferential solvation at the inner edge of excess hydration. *J. Chem. Phys.*, *146*, 215101-8.
[37] Sirotkin, V. A. & Kuchierskaya, A. A. (2017). α-Chymotrypsin in water-ethanol mixtures: Effect of preferential interactions. *Chem. Phys. Lett.*, *689*, 156-161.
[38] Sirotkin, V. A. & Kuchierskaya, A. A. (2017). α-Chymotrypsin in water-acetone and water-dimethyl sulfoxide mixtures: Effect of preferential solvation and hydration. *Proteins: Functions, Structure and Bioinformatics*, *85*, 1808-1819.

Chapter 2

PREFERENTIAL SOLVATION OF α-CHYMOTRYPSIN IN WATER-MONOHYDRIC ALCOHOL MIXTURES: EFFECT OF CHAIN LENGTH

Vladimir A. Sirotkin and Alexandra A. Kuchierskaya
A. M. Butlerov Institute of Chemistry,
Kazan Federal University, Kazan, Russia

ABSTRACT

Preferential solvation and hydration of bovine pancreatic α-chymotrypsin was investigated in the entire range of water content in monohydric alcohols at 25°C. This approach is based on the analysis of absolute values of the water and monohydric alcohol sorption. One of the most important advantages of our approach is the facilitation of individual evaluation of the Gibbs energies of water, alcohol, and protein in the entire range of water content. This methodology was applied to estimate protein destabilization/stabilization of lysozyme in water-monohydric alcohol mixtures.

Three distinct schemes are operative in water-alcohol mixtures. α-chymotrypsin is preferentially hydrated at high water content. Protein has a higher affinity for alcohol than for water at intermediate water content. At low water content, preferential solvation of chymotrypsin depends significantly on the alkyl chain length. Opposite to methanol, ethanol and propanol-1 are preferentially excluded from the protein surface at the lowest water content. This results in preferential hydration of α-chymotrypsin in the water-poor ethanol and propanol-1. Our data clearly show that the length of the alkyl chain of monohydric alcohols is one of the critical factors in determining the stability of protein-water-alcohol systems.

Keywords: preferential solvation, preferential hydration, α-chymotrypsin, water, alcohol, ethanol

1. INTRODUCTION

It is well-known that the stability, structure, and functions of globular proteins are governed by interactions of the protein macromolecules with water [1-15]. Small monohydric alcohols (methanol, ethanol, and propanol-1) are widely utilized in biophysical, biomedical, and biotechnological investigations to change the protein-water interactions and consequently modulate the protein stability. In particular, there are many advantages in employing water-poor alcohols, including the suppression of undesirable side reactions caused by water, the biocatalysis of reversed hydrolytic reactions (transesterification), or increased thermostability [16-31]. Distinct intermediate protein states induced by alcohols may be responsible for numerous neurodegenerative diseases (Alzheimer's disease, Parkinson's disease, and Huntington's disease) [32-35]. However, the manner in which alcohols increase/decrease the thermal stability, induce/reduce the extent of denaturation, and stabilize/destabilize the partially folded conformations of proteins (amyloid fibrils and molten globules) is an intricate function of water content in alcohols.

There are numerous investigations which demonstrate the bilateral action of monohydric alcohols on the protein properties [36-41]. For example, the temperature of protein denaturation in monohydric alcohols decreases gradually with augmenting organic solvent concentration [36-39]. This effect becomes more pronounced with increasing length of the alkyl chain. On the other hand, the denaturation enthalpy passes through a maximum with augmenting alcohol content [37-41]. At low alcohol content and temperatures around 0-25°C, monohydric alcohols can slightly stabilize proteins [39-41].

Understanding the bilateral impact of monohydric alcohols on the protein stability requires effective techniques that reveal biophysical information regarding protein–alcohol and protein–water interactions. Preferential solvation/hydration may be an effective and informative approach for elucidating the dual effect of water-alcohol mixtures on the protein stability. Preferential solvation is a thermodynamic quantity that describes the protein occupancy by the alcohol/water molecules [42-57]. Alcohol and water exist preferentially in the solvation layer of the protein. When a protein is placed into a water-alcohol mixture, its properties are altered as a function of the solvent composition. The preferential solvation/hydration process accounts for the augmentation or depletion of the alcohol/water molecules at the protein surface. Preferential binding is the excess of alcohol at the protein surface relative to the alcohol content in the bulk solvent. The preferential binding depends markedly on the chemical nature of the protein surface. For example, protein unfolding may be induced by the preferential binding to specific regions on the protein (to peptide groups in the case of urea and guanidinium hydrochloride or to hydrophobic regions in the case of alcohols) [42, 43, 47-57].

The aim of our study is to monitor the preferential solvation and preferential hydration of the protein macromolecules at low, intermediate, and high water content in monohydric alcohols at 25°C. Our approach is based on a simultaneous analysis of residual enzyme activity and absolute values of the water/alcohol sorption. One of the most important advantages of our approach is the facilitation of individual evaluation of the Gibbs energies of water, alcohol, and protein in the entire range of water content.

Bovine pancreatic α-chymotrypsin (CT) was used as a model protein. It is one of the most studied and applied in the protein science [58, 59]. The physiological role of CT is to hydrolyze peptide bonds [58, 59]. α-chymotrypsin is an example of a predominantly β-sheet protein.

The choice of methanol, ethanol, and propanol-1 was determined for the following reasons:

a) They are widely utilized in enzymology and biotechnology.
b) They are organic solvents that are water-miscible. Therefore, it is possible to investigate the effects of these organic substances on the hydration and enzyme activity of α-chymotrypsin in the entire range of water content.
c) These organic solvents represent a series of liquids in which the size of hydrophobic chain is gradually increased when going from methanol to propanol-1.

This means that these organic liquids may be considered as an informative tool for analyzing the effect of length of the alkyl chain on preferential interactions in the protein-water-alcohol systems.

2. Experimental

2.1. Materials

Bovine pancreatic α-chymotrypsin was purchased from Sigma Chemical Co. (St. Louis, MO, U.S.A., No. C 4129, protein content >95%; essentially salt free; EC 3.4.21.1; 66 units of specific activity for N-benzoyl-L-tyrosine ethyl ester). The molecular mass of bovine pancreatic α-chymotrypsin was determined to be 25000 Da. Methanol,

ethanol, and propanol-1 (analytical grade, with a purity of > 99%) were purified and dried according to the recommended guidelines [60]. Water used was doubly distilled. All water-organic mixtures were prepared gravimetrically using a Precisa balance (Swiss) with a precision of 0.00001 g.

2.2. Initial Protein State

The α-chymotrypsin powder was placed on the thermostated cell and dried using a microthermoanalyzer "Setaram" MGDTD-17S (±0.00001 g) at 25°C and 0.1 Pa, until a constant sample weight was reached. The water content of the dehydrated protein was estimated as 0.002 ± 0.001 g water/g protein, using the Karl Fischer titration method as described previously [61-64]. This value for lysozyme implies that at the **zero hydration level** there are about three water molecules strongly bound to each protein molecule.

2.3. Sorption Measurements

The protein samples were prepared according to the recommended guidelines [61, 62]. The initially dried protein samples were presented to water-alcohol vapor mixtures. The water-organic vapor mixture was consecutively flowed through a thermostated saturator filled with the water-alcohol mixture, and a cell containing the protein sample. Protein samples (5-10 mg) each were flushed by water – alcohol vapor mixtures, until no further mass changes were detected as described previously. The sorption equilibrium was reached after 6 h at 25°C. The schematic representation of the experimental setup is given in Ref. 61. An external ethylene glycol thermostat (RC 6, Lauda, Germany) was utilized to determine the temperature with a precision of 0.1 °C. The water activity (a_1) in the vapor phase was adjusted by altering the water content in the liquid water-alcohol mixture.

Measurements of the water sorption (A_1) were conducted by Karl Fischer titration with a Metrohm 831 KF coulometer. Alcohol content of lysozyme (A_3) was calculated as a difference between the total sorption uptake (A_1+A_3) and water content (A_1). The total sorption uptake (A_1+A_3) was measured by microthermoanalyzer "Setaram" MGDTD-17S.

2.4. FTIR Spectroscopy

FTIR spectra were recorded with a Nicolet MAGNA 550 infrared spectrometer. FTIR spectra were measured at 25°C as described previously [61]. Co-adding 256 scans at a spectral resolution of 2 cm^{-1} achieved each spectrum. The sample chamber was purified with dry and carbon dioxide free air. Glassy-like protein films casted from 2% (w/v) water solution onto the CaF_2 window at room humidity and temperature achieved the infrared spectra. A schematic representation of the experimental conditions for FTIR spectroscopic measurements of the proteins' structural characteristics is given in Ref. 61. Once the window was mounted in the sample cell, the film was dehydrated through flushing air that was dried over the P_2O_5 powder. At 25°C, relative water vapor pressure (water activity, a_w) over P_2O_5 did not exceed 0.01 [65].

The protein films were flushed until there was no detection of additional spectral changes in the 3450 cm^{-1} water absorbance region, as well as a smooth line without any visible shoulders was represented by a contour on this side. The spectrum of this sample was utilized as a reference spectrum for the calculation of the difference spectra. The initially dehydrated protein was exposed to water or organic solvents.

2.5. Analysis and Band Assignment of Protein Infrared Spectra

The infrared spectra of proteins contain some vibrational bands arising from the amide groups of in the protein backbone. Among these, the amide I band between 1700 and 1600 cm^{-1}, which is due primarily to the C=O stretch of the peptide linkages, is the most sensitive to changes in the secondary structure of proteins. The infrared spectra of α-chymotrypsin in the absence of organic solvents in the amide I region are presented in Figures 1 and 2. Figures 1 and 2 show typical absorbance and second derivative spectra of α-chymotrypsin in the amide I region.

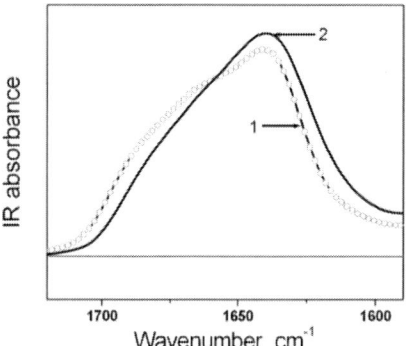

Figure 1. Typical absorbance spectra of α-chymotrypsin in the amide I region in the absence of organic solvents: (1) a_w=0.01; (2) pure water.

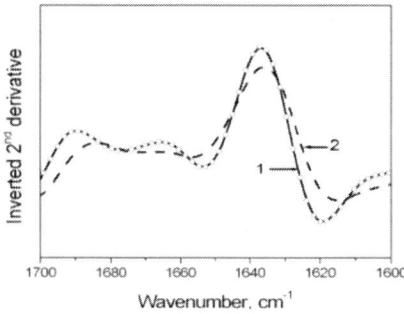

Figure 2. Typical second derivative spectra of α-chymotrypsin in the amide I region in the absence of organic solvents: (1) a_w=0.01; (2) pure water.

The amide I contour consists of a number of bands at frequencies that are characteristic of specific types of secondary structure. It has been well documented that the frequency at the maximum absorbance of the amide I band is determined by the predominant form of secondary structure in the protein [61, 66-68]. The maximum absorbance of the amide I band of proteins with a high content of α-helix occurs between 1658 and 1654 cm^{-1} and with a high content of β-sheet structure the maximum is found between 1643 and 1632 cm^{-1}.

As concluded from Figures 1 and 2, the most dominant band component of the chymotrypsin spectra is the band at 1637 cm^{-1}, which is attributed to a β-sheet structure [66-69]. The band at 1685 cm^{-1} was assigned to an intermolecular β-sheet structure [66-69].A minor component at 1670-1665 cm^{-1} was assigned to irregular secondary structures (β-turns and extended chains) [66-69].

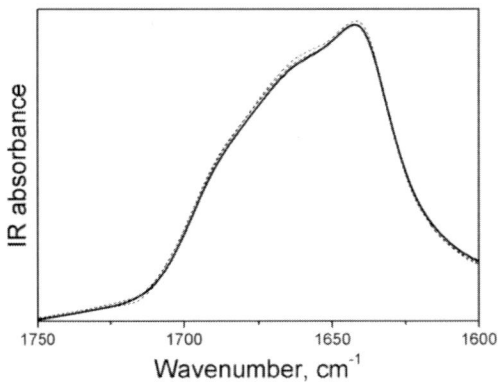

Figure 3. Amide I spectra of the dried chymotrypsin: Solid line – in the absence of organic solvent. Dashed line – in neat propanol-1.

Figure 3 shows the α-chymotrypsin spectra in neat propanol-1. No noticeable changes compared with the spectrum of the dried chymotrypsin were observed in this case. Considerable changes were observed in anhydrous methanol (Figure 4). The interaction of the dried chymotrypsin with methanol results not only in generating some helical structure but also in the subsequent formation of more extensive β-sheets (Figure 4).

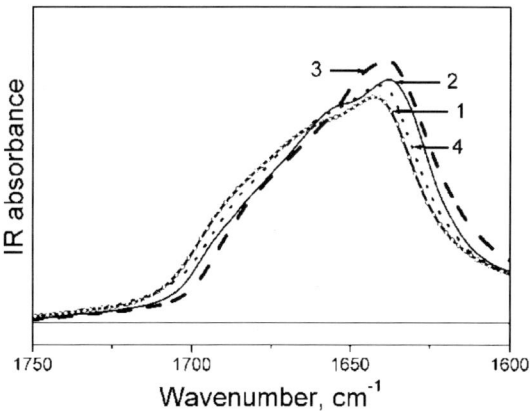

Figure 4. Amide I spectra of the dried chymotrypsin: (1) in the absence of organic solvent; (2) Methanol; (3) Pure water; (4) Ethanol.

2.6. Enzyme Activity

Residual enzyme activity was determined by measuring the enzyme activity after storage in water - alcohol mixtures as described previously [64]. The model process was the hydrolysis of N-acetyl-L-tyrosine ethyl ester (ATEE) catalyzed by α-chymotrypsin. The measurements were performed on a TitroLine Alpha Plus potentiometric titrator (Schott Instruments, Germany) in the pH-static mode at pH 8.0 and 25°C. The concentration of the substrate was 4.0×10^{-3} mol/L. In the course of the experiment, the pH value was maintained at a constant level by adding a titrant (a potassium hydroxide solution of known concentration), which neutralized the acid (N-acetyl-L-tyrosine) released during the hydrolysis. The kinetic curve obtained was the time dependence of the amount of the reagent spent for titrating the acid released. Each kinetic curve was reproduced not less than three times.

The reaction mixture was prepared as follows. The initially dehydrated α-chymotrypsin was immersed in an aqueous-organic mixture of required composition, and was incubated at 25°C for 180 min. The concentration of the α-chymotrypsin in the water-alcohol mixture was 1 mg/ml. Adding 100-μl aliquots of the α-chymotrypsin solution in the water-rich organic solvents (or the α-chymotrypsin suspension in the water-poor alcohols) to the aqueous solution of the substrate, we initiated the enzymatic reaction.

3. RESULTS AND DISCUSSION

3.1. Analysis of Sorption Isotherms

Figures 5 and 6 present the water (A_1) and alcohol (A_3) vapor sorption isotherms for α-chymotrypsin at 25°C. The protein solvation layer is composed of two parts: (i) nonideal (due to preferential interactions) and (ii) ideal. The nonideal effect of the solvation layer on enzyme properties (alcohol/water sorption; residual enzyme activity) can be defined in terms of excess functions, F^E [70-72], which refers to the difference between the observed function of mixing, F^M, and the function for an ideal binary mixture, F_{id}^M.

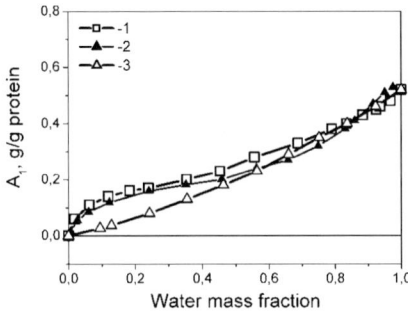

Figure 5. Water (A_1) sorption isotherms for α-chymotrypsin at 25°C: 1 – Propanol-1; 2 – Ethanol; 3 – Methanol.

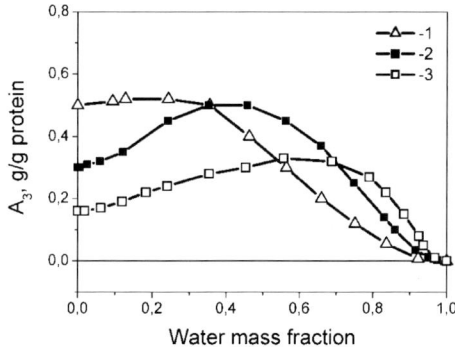

Figure 6. Alcohol (A_3) sorption isotherms for α-chymotrypsin at 25°C: 1 – Methanol; 2 – Ethanol; 3 – Propanol-1.

The deviations of the excess functions from zero describe the degree to which the protein solvation layer differs from the pure binary water-alcohol mixture as a consequence of preferential interactions between water (component 1), protein (component 2), and alcohol (component 3).

Eq. 1 is used to calculate the F^E values:

$$F^E = F^M - F^M_{id} \qquad (1)$$

The $F^M_{id,i}$ values were calculated using Eq. 2:

$$F^M_{id,i} = F^M_i(w_i = 0) + w_i[F^M_i(w_i = 1.0) - F^M_i(w_i = 0)] \qquad (2)$$

where $F^M_i(w_i = 1.0)$ is the observed mixing function of α-chymotrypsin at $w_i = 1.0$; $F^M_i(w_i = 0)$ is the observed mixing function of α-chymotrypsin at $w_i = 0$; w_1 is the water mass fraction in the binary water-organic mixtures; and w_3 is the organic solvent mass fraction in the binary water-organic mixtures ($w_1 + w_3 = 1.0$).

The situation in which no preferential interactions exist between water, protein, and alcohol is described by the $F^M_{id,i}$ values. In this circumstance, the water mass fraction in the ideal part of the solvation layer is the same as in the pure water-alcohol mixture.

The Z_1^M (water mass fraction in the solvation layer) and Z_3^M (alcohol mass fraction in the solvation layer) values as a function of water mass fraction in alcohol are presented in Figure 7. Eqs. 3 and 4 are utilized to calculate the Z_1^M and Z_3^M values, respectively:

$$Z_1^M = \frac{A_1}{A_1+A_3} \qquad (3)$$

$$Z_3^M = \frac{A_3}{A_1+A_3} \qquad (4)$$

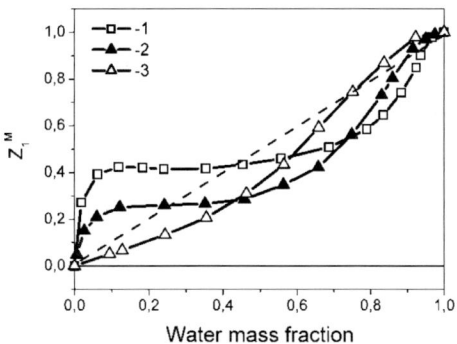

Figure 7. Water mass fraction in the solvation layer of α-chymotrypsin (Z_1^M): 1 – Propanol-1; 2 – Ethanol; 3 – Methanol.

The simultaneous action of alcohol and water was determined by the Z_1^E and Z_3^E values (Figure 8). These excess sorption functions can be estimated using Eqs. 5 and 6:

$$Z_1^E = Z_1^M - Z_{id,1}^M \qquad (5)$$

$$Z_3^E = Z_3^M - Z_{id,3}^M \qquad (6)$$

where Z_1^M is the mass fraction of water in the solvation layer for the real water-alcohol mixture; and $Z_{id,1}^M$ is the mass fraction of water in the solvation layer for the ideal water-alcohol mixture. The $Z_{id,1}^M$ values were calculated using Eq. 7:

$$Z_{id,1}^M = Z_1^M(w_1 = 0) + w_1[Z_1^M(w_1 = 1.0) - Z_1^M(w_1 = 0)] \qquad (7)$$

where $Z_1^M(w_1 = 1.0)$ is the water mass fraction in the solvation layer of α-chymotrypsin at w_1=1.0; $Z_1^M(w_1 = 0)$ is the water mass fraction in the solvation shell at w_1=0; and w_1 is the mass fraction of water in alcohol.

In addition, Z_3^M is the alcohol mass fraction in the solvation layer for the real water-alcohol mixture; and $Z_{id,3}^M$ is the alcohol mass fraction for the ideal water-alcohol mixture. Eq. 8 is used to calculate the $Z_{id,3}^M$ values:

$$Z_{id,3}^M = Z_3^M(w_3 = 0) + w_3[Z_3^M(w_3 = 1.0) - Z_3^M(w_3 = 0)] \qquad (8)$$

where $Z_3^M(w_3 = 0)$ is the alcohol mass fraction in the solvation shell of α-chymotrypsin at w_3=1.0; $Z_3^M(w_3 = 0)$ is the alcohol mass fraction in the solvation shell of lysozyme at w_3=0; and w_3 is the alcohol mass fraction in the binary water-organic mixture.

3.1.1. Propanol-1 and Ethanol

As shown in Figure 8, the Z_1^E values are positive at high (w_1 ~ 0.9-1.0) and low (w_1 ~ 0-0.3) water content. A considerable decrease in the water sorption was observed in the intermediate range of water content. The Z_1^E values are negative in this concentration region. The most pronounced decline was found in the water mass fraction range from 0.4 to 0.8.

3.1.2. Methanol

Figure 8 shows that the Z_1^E values are positive at high ($w_1 = 0.7$-1.0) water content. The Z_1^E values are negative at low water content ($w_1 = 0$-0.5).

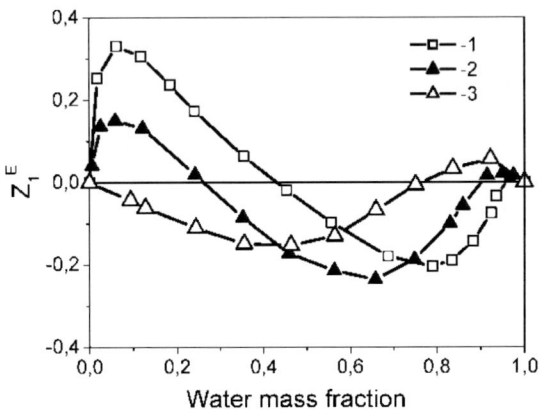

Figure 8. Excess water mass fraction in the solvation layer of α-chymotrypsin in water-alcohol mixtures (Z_1^E): 1 – Propanol-1; 2 – Ethanol; 3 – Methanol.

3.2. Residual Enzyme Activity

Figure 9 demonstrates typical kinetic curves for the enzymatic reaction catalyzed by α-chymotrypsin preliminarily incubated in water-alcohol mixtures. The catalytic activity was characterized by the ratio of the extent of hydrolysis attained within 200 s with α-chymotrypsin incubated in a water-alcohol mixture to the same quantity measured using CT incubated in pure water (Figure 9, curve 1).

The residual activity values are shown in Figure 10. As shown in Figure 10, alcohols affect the residual enzyme activity of α-chymotrypsin in a complicated manner.

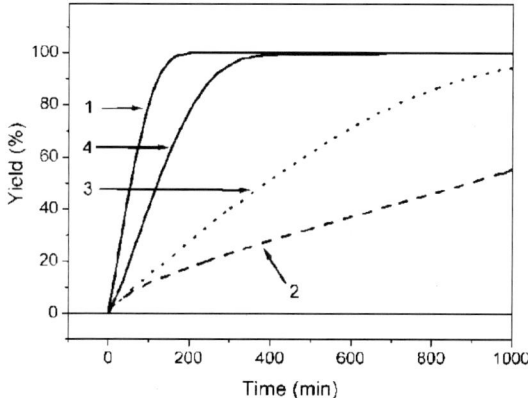

Figure 9. Typical kinetic curves for the enzymatic reaction catalyzed by α-chymotrypsin previously incubated in water-alcohol mixtures. Propanol-1. Water mass fraction: (1) 1.0; (2) 0.84; (3) 0.55; (4) 0.01.

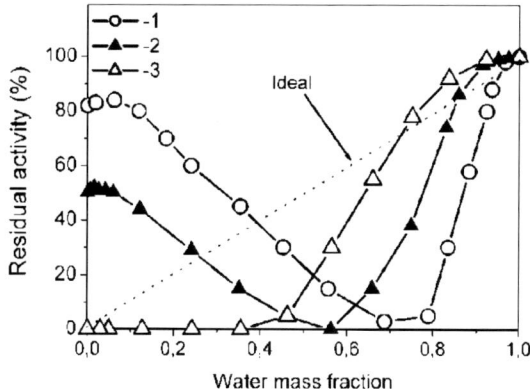

Figure 10. (A) Residual activity of the dried α-chymotrypsin in water-alcohol mixtures: 1 – Propanol-1; 2 – Ethanol; 3 - Methanol. (B) Residual activity of the hydrated α-chymotrypsin in water-alcohol mixtures: 1 – Propanol-1; 2 – Ethanol; 3 - Methanol.

3.2.1. Propanol-1 and Ethanol

The residual activity values are close to 100% (Figure 10) at high water content ($w_1 \sim 0.9$-1.0). There is a sharp transition from the water-rich range to the intermediate one at $w_1 < 0.9$. The residual enzyme activity of CT changes from 100 to 0% in the transition region ($w_1 \sim 0.8$-0.9 for propanol-1; $w_1 \sim 0.6$-0.8 for ethanol). A minimum on the residual activity curve was found at $w_1 \sim 0.8$ in propanol-1 and at $w_1 \sim 0.6$ in ethanol. The residual activity augments at $w_1 < 0.4$. At low water content, the residual enzyme activity is ~50% (in water-poor ethanol) and ~80% (in water-poor propanol-1), compared with that observed after incubation in pure water.

3.2.2. Methanol

The residual activity values are close to 100% (Figure 10), at high water content ($w_1 \sim 0.8$-1.0). The residual enzyme activity alters from 100 to 0% in the transition range ($w_1 \sim 0.2$-0.8). At low water content, opposite to ethanol and propanol-1, no residual activity was observed.

3.3. Excess Residual Enzyme Activity

The effect of the excess hydration (Z_1^E) of the residual enzyme activity was described by the R_1^E values (excess residual enzyme activity) (Figure 11). The R_1^E values were estimated using Eq. 9:

$$R_1^E = R^M - R_{id,1}^M \qquad (9)$$

where R^M is the observed residual enzyme activity; and $R_{id,1}^M$ is the function for an ideal binary mixture.

The R_{id}^M values can be calculated using Eq. 10:

$$R_{id,1}^M = R^M(w_1 = 0) + w_1[R^M(w_1 = 1.0) - R^M(w_1 = 0)] \qquad (10)$$

where $R^M(w_1 = 1.0)$ is the residual activity of α-chymotrypsin at $w_1 = 1.0$; $R^M(w_1 = 0)$ is the residual activity of α-chymotrypsin at $w_1 = 0$; w_1 is the water mass fraction in the binary water-alcohol mixtures; and w_3 is the alcohol mass fraction in the binary water-organic mixtures ($w_1 + w_3 = 1.0$).

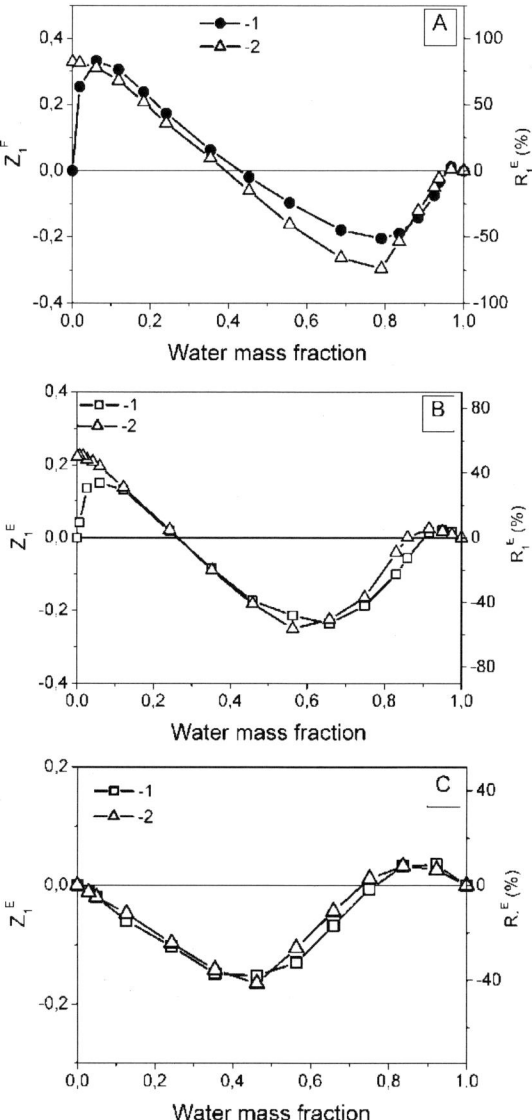

Figure 11. (1) Excess water mass fraction in the solvation layer of the dried α-chymotrypsin (Z_1^E). (2) Excess residual activity of α-chymotrypsin in water-alcohol mixtures (R_1^E). A – Propanol-1; B – Ethanol; C - Methanol.

Figure 11 demonstrates the dependencies of the R_1^E values on the water content in alcohols. The average W-S interactions in the solvation layer are the same as the average S-S and W-W interactions in the bulk solvent in ideal binary mixtures (mixtures of two components, S [alcohol] and W [water]). Nonideal mixtures comprise molecules for which the S-S, W-W, and W-S interactions are all distinct. As concluded from Figure 11, the R_1^E values differ significantly from zero. This means that the effect of the water-alcohol solvation shell on the residual enzyme activity is nonideal in the entire range of water content. It is worth noting that the R_1^E values are consistent with the Z_1^E values (Figure 11).

Three concentration regions were observed for ethanol and propanol-1 (Figure 11):

i. At high water content (Figure 4), the R_1^E and Z_1^E values are positive.
ii. At intermediate water content, the R_1^E and Z_1^E values are negative. Ethanol and propanol-1 augment the degree of irreversible inactivation of α-chymotrypsin in this region.
iii. At low water content, the R_1^E and Z_1^E values are positive.

Two different regimes are operative for methanol (Figure 11):

i. The $\boldsymbol{R_1^E}$ and $\boldsymbol{Z_1^E}$ values are positive at intermediate and high water content (Figure 5).
ii. At the lowest water content, the $\boldsymbol{R_1^E}$ and $\boldsymbol{Z_1^E}$ values are negative. In contrast to ethanol and propanol-1, the alcohol-induced irreversible inactivation was detected at low water content for methanol.

3.4. Preferential Solvation Parameters

The degree to which the solvation layer differs from the pure binary water-alcohol mixture due to preferential interactions between water (component 1), protein (component 2), and alcohol (component 3) can be characterized by the preferential interaction parameters (Eqs. 11 and

12). Eq. 11 [42-44] was utilized to determine the preferential binding of lysozyme in water-alcohol mixtures:

$$(\partial g_3/\partial g_2)_{T,\mu_1,\mu_3} = A_3 - \frac{w_3}{w_1} A_1 \qquad (11)$$

where A_1 is the protein hydration, expressed as gram water per gram protein; A_3 is the binding of alcohol, expressed as gram organic solvent per gram protein; w_1 is the water mass fraction in water-alcohol mixtures; and w_3 is the mass fraction of organic solvent in water-alcohol mixtures ($w_1 + w_3 = 1.0$).

Eq. 12 was used to estimate the preferential hydration:

$$(\partial g_1/\partial g_2)_{T,\mu_1,\mu_3} = -(\frac{w_1}{w_3})(\partial g_3/\partial g_2)_{T,\mu_1,\mu_3} \qquad (12)$$

Figures 12 and 14 show the preferential interaction parameters calculated using Eqs. 11 and 12, respectively.

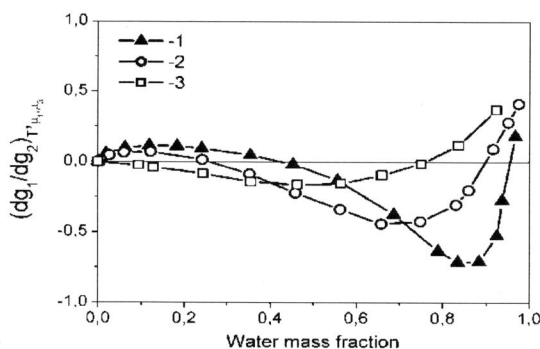

Figure 12. The preferential hydration parameters of α-chymotrypsin as a function of water mass fraction in alcohols (($\partial g_1/\partial g_2)_{T,\mu_1,\mu_3}$): 1 – Propanol-1; 2 – Ethanol; 3 – Methanol.

Eqs. 13 and 14 were utilized to determine the Gibbs energies of the transfer of water (ΔG_1^{pref}) and alcohol (ΔG_3^{pref}) from water-alcohol mixtures to the solvation layer of α-chymotrypsin:

$$\Delta G_1^{pref} = \mu_1^E(solvation\ layer) - \mu_1^E(binary\ mixture) \qquad (13)$$

$$\Delta G_3^{pref} = \mu_3^E(solvation\ layer) - \mu_3^E(binary\ mixture) \qquad (14)$$

The $\mu_1^E(binary\ mixture)$ and $\mu_3^E(binary\ mixture)$ values can be calculated using Eqs. 15 and 16:

$$\mu_1^E(binary\ mixture) = RT\ln\gamma_1(binary\ mixture) \qquad (15)$$

$$\mu_3^E(binary\ mixture) = RT\ln\gamma_3(binary\ mixture) \qquad (16)$$

Water activity coefficients ($\gamma_1(binary\ mixture)$, the mass fraction scale; the reference state is pure water) in aqueous-organic mixtures were estimated using Eq. 17:

$$\gamma_1(binary\ mixture) = \frac{a_1}{w_1} \qquad (17)$$

Alcohol activity coefficients ($\gamma_3(binary\ mixture)$, the mass fraction scale; the reference state is pure alcohol) in aqueous-organic mixtures were calculated using Eq. 18:

$$\gamma_3(binary\ mixture) = \frac{a_3}{w_3} \qquad (18)$$

Water activity (a_1) and alcohol activity (a_3) were taken from the published data [73] based on the vapor-liquid equilibrium.

The $\mu_1^E(solvation\ layer)$ and $\mu_3^E(solvation\ layer)$ values were calculated using Eqs. 19 and 20:

$$\mu_1^E(solvation\ layer) = RT\ln\gamma_1(solvation\ layer) \qquad (19)$$

$$\mu_3^E(solvation\ layer) = RT\ln\gamma_3(solvation\ layer) \qquad (20)$$

Water activity coefficients ($\gamma_1(solvation\ layer)$, the mass fraction scale) in the solvation layer were calculated using Eq. 21:

$$\gamma_1(solvation\ layer) = \frac{a_1}{Z_1^M} \quad (21)$$

where Z_1^M is the mass fraction of water in the solvation layer;

$$Z_1^M = \frac{A_1}{A_1+A_3}.$$

Alcohol activity coefficients ($\gamma_3(solvation\ layer)$, the mass fraction scale) in the solvation layer were estimated using Eq. 22:

$$\gamma_3(solvation\ layer) = \frac{a_3}{Z_3^M} \quad (22)$$

where Z_3^M is the alcohol mass fraction in the solvation layer;

$$Z_3^M = \frac{A_3}{A_1+A_3}.$$

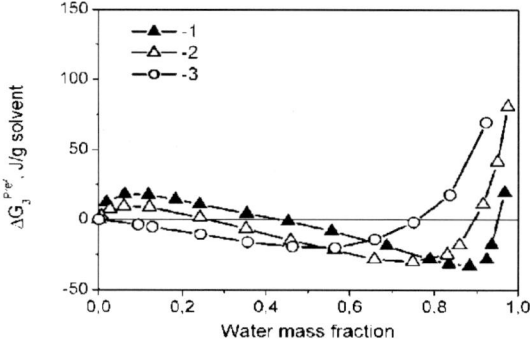

Figure 13. Gibbs energy of the transfer of alcohols (ΔG_3^{pref}) from binary water-organic mixtures to the solvation shell of α-chymotrypsin: 1 – Propanol-1; 2 – Ethanol; 3 – Methanol.

Figures 13 and 15 show the ΔG_3^{pref} and ΔG_1^{pref} values. Figures 9-12 demonstrate that the ΔG_3^{pref} and ΔG_1^{pref} values correlate well with the preferential interaction parameters.

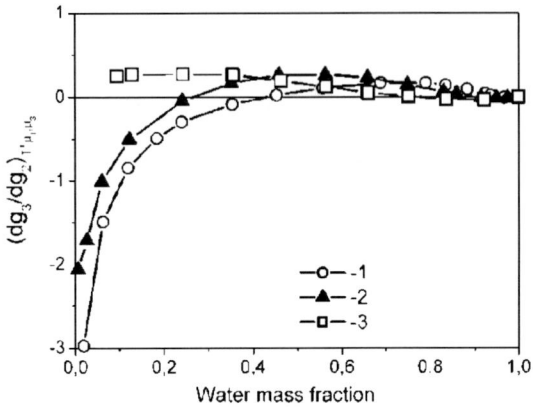

Figure 14. The preferential solvation parameters of α-chymotrypsin as a function of water mass fraction in alcohols $((\partial g_3/\partial g_2)_{T,\mu_1,\mu_3})$: 1 – Propanol-1; 2 – Ethanol; 3 – Methanol.

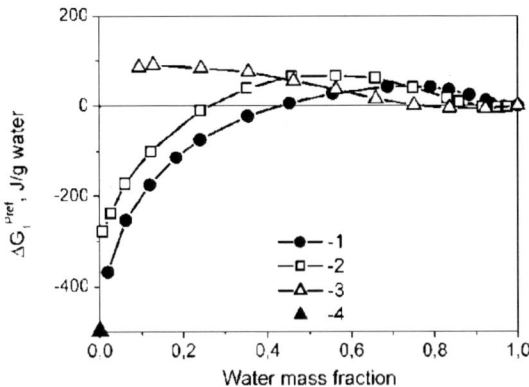

Figure 15. Gibbs energy of the transfer of water (ΔG_1^{pref}) from binary water-organic mixtures to the solvation shell of α-chymotrypsin: 1 – Propanol-1; 2 – Ethanol; 3 – Methanol. 4 - Excess Gibbs energy of water bound to the dried protein [6]. Reference state is pure liquid water at 25°C.

The ΔG_1^{pref} value at w_1=0 (Figure 15) was compared with the excess Gibbs energy of water bound to the dried protein [6]. This Gibbs energy was determined from the water sorption experiments without propanol-1. As shown in Figure 15, strong agreement between our data and previously published results was observed. This fact confirms the reliability of our calculations.

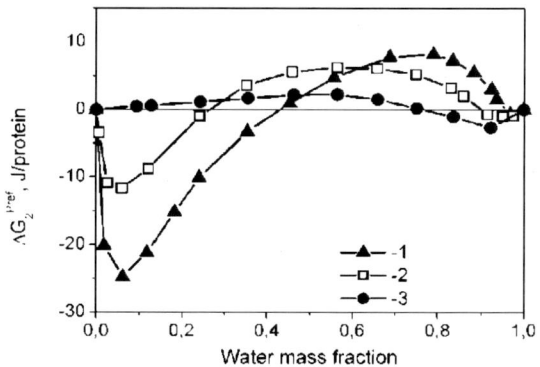

Figure 16. Gibbs energy (ΔG_2^{pref}) of the transfer of α-chymotrypsin from pure state to water-alcohol mixtures: 1 – Propanol-1; 2 – Ethanol; 3 – Methanol.

ΔG_2^{pref} values (Figure 16) described α-chymotrypsin stabilization and destabilization due to the preferential interactions. The Gibbs-Duhem equation for ternary systems (Eq. 23) was utilized to calculate the ΔG_2^{pref} values:

$$\Delta G_2^{pref} = \frac{m_1 \Delta G_1^{pref} + m_3 \Delta G_3^{pref}}{m_2} \tag{23}$$

$$\Delta G_2^{pref} = \mu_2^E(protein\ in\ water - alcohol\ mixtures) - \mu_2^E(pure\ protein) \tag{24}$$

where μ_1^E, μ_2^E, and μ_3^E are the excess chemical potentials of water, protein, and alcohol, respectively; and m_1, m_2, and m_3 are the masses of water, protein, and alcohol, respectively.

3.5. Effect of Length of the Alkyl Chain on Preferential Interactions

The enzyme activity and sorption experiments can be summarized as follows. Three regimes were found for α-chymotrypsin:

i. α-chymotrypsin is in the native (preferentially hydrated) state at high water content. α-chymotrypsin has a higher affinity for water than for monohydric alcohols. The ΔG_1^{pref}, ΔG_2^{pref}, and $(\partial g_3/\partial g_2)_{T,\mu_1,\mu_3}$ values are negative. On the other hand, the $(\partial g_1/\partial g_2)_{T,\mu_1,\mu_3}$ and Z_1^E values are positive.
ii. The preferential binding of alcohols to α-chymotrypsin was observed at intermediate water content. The $(\partial g_3/\partial g_2)_{T,\mu_1,\mu_3}$, ΔG_1^{pref}, and ΔG_2^{pref} values are positive. On the other hand, the $(\partial g_1/\partial g_2)_{T,\mu_1,\mu_3}$ and Z_1^E values are negative. Alcohol-induced irreversible inactivation was found in this concentration range [63].
iii. The deep dehydration of α-chymotrypsin leads to the formation of a rigid (glassy-like) state as a consequence of the proton-transfer phenomena and hydrogen bonding between ionizable and polar protein residues. [1-3]. The bulky propanol-1 molecules are incapable of breaking the dehydration-induced protein-protein contacts by themselves, in the absence of water, likely because of steric and diffusion limitations. The molar volume of propanol-1 (75.1 cm³/mol) is more than four times larger than that of water (18 cm³/mol). Therefore, no significant structural rearrangements were observed for propanol-1 (Figure 3).

Hence, it is expected that bulky propanol-1 molecules are not effective in solvating the dehydrated protein alone. This indicates that the propanol-1 molecules are preferentially excluded from the enzyme surface. It results in the preferential hydration of α-chymotrypsin.

Therefore, the $(\partial g_1/\partial g_2)_{T,\mu_1,\mu_3}$ values are positive at low water content.

On the other hand, methanol (The molar volume is 41.0 cm^3/mol.) is effective in solvating the dehydrated protein alone, in the absence of water. It results in preferential solvation of α-chymotrypsin by methanol molecules. Therefore, considerable non-native structural rearrangements were observed in methanol (Figure 4). Ethanol (The molar volume is 58.6 cm^3/mol) shows intermediate behavior (Figure 4).

Our data clearly demonstrate that the length of the alkyl chain of monohydric alcohols is one of the critical factors in determining the stability of protein-water-alcohol systems.

REFERENCES

[1] Gregory, R. B. (1995). Protein hydration and glass transition behavior. In: *Protein-Solvent Interactions* (Gregory, R. B., ed.), Marcel Dekker, New York, 191-264.

[2] Rupley, J. A. & Careri, G. (1991). Protein hydration and function. *Adv. Protein Chem.*, *41*, 37-172.

[3] Kuntz, I. D. & Kauzmann, W. (1974). Hydration of proteins and polypeptides. *Adv. Protein Chem.*, 28, 239-345.

[4] Oleinikova, A., Smolin, N., Brovchenko, I., Geiger, A. & Winter, R. (2005). Properties of spanning water networks at protein surfaces. *J. Phys. Chem. B.*, *109*, 1988-1998.

[5] Durchschlag, H. & Zipper, P. (2001). Comparative investigations of biopolymer hydration by physicochemical and modeling techniques. *Biophys. Chem.*, *93*, 141-157.

[6] Sirotkin, V. A. & Khadiullina, A. V. (2013). Gibbs energies, enthalpies, and entropies of water and lysozyme at the inner edge of excess hydration. *J. Chem. Phys.*, *139*, 075102/1-075102/9.

[7] Sirotkin, V. A. & Khadiullina, A. V. (2014). A study of the hydration of ribonuclease A using densitometry: Effect of the protein hydrophobicity and polarity. *Chem. Phys. Lett*, *603*, 13-17.

[8] Privalov, P. L. & Crane-Robinson, C. (2017). Role of water in the formation of macromolecular structures. *Eur. Biophys. J.*, *46*, 203-224.

[9] Bull, H. B. (1944). Adsorption of water vapor by proteins. *J. Amer. Chem. Soc.*, 66, 1499-1507.

[10] Luscher-Mattli, M. & Ruegg, M. (1982). Thermodynamic functions of biopolymer hydration. I. Their determination by vapor pressure studies, discussed in an analysis of the primary hydration process. *Biopolymers*, 21, 403-418.

[11] Luscher-Mattli, M. & Ruegg, M. (1982). Thermodynamic functions of biopolymer hydration. II. Enthalpy-entropy compensation in hydrophilic hydration process. *Biopolymers*, 21, 419-429.

[12] Bone, S. (1987). Time-domain reflectometry studies of water binding and structural flexibility in chymotrypsin. *Biochem. Biophys. Acta.*, *916*, 128-134.

[13] Sirotkin, V. A. & Khadiullina, A. V. (2011). Hydration of proteins: excess partial enthalpies of water and proteins. *J. Phys. Chem. B*, *115*, 15110-15118.

[14] Sirotkin, V. A., Komissarov, I. A. & Khadiullina, A. V. (2012). Hydration of proteins: excess partial volumes of water and proteins, *J. Phys. Chem. B*, *116*, 4098-4105.

[15] Sirotkin, V. A. (2005). Effect of dioxane on the structure and hydration-dehydration of α-chymotrypsin as measured by FTIR spectroscopy. *Biochim. Biophys. Acta.*, *1750*, 17-29.

[16] Klibanov, A. M. (2001). Improving enzymes by using them in organic solvents, *Nature*, *409*, 241-246.

[17] Carrea, G. & Riva, S. (2000). Properties and synthetic applications of enzymes in organic solvents, *Angew. Chem. Int. Ed.*, *39*, 2226-2254.

[18] Halling, P. J. (2004). What can we learn by studying enzymes in nonaqueous media? *Phil. Trans. R. Soc. Lond. B Biol. Sci.*, *359*, 1287–1297.

[19] Micaelo, N. M. & Soares C. M. (2007). Modeling hydration mechanisms of enzymes in nonpolar and polar organic solvents, *FEBS J.*, *274*, 2424–2436.

[20] Clark, D. S. (2004). Characteristics of nearly dry enzymes in organic solvents: implications for biocatalysis in the absence of water, *Phil. Trans. R. Soc. Lond. B Biol. Sci.*, *359*, 1299–1307.

[21] Serdakowski, A. L. & Dordick, J. S. (2007). Enzyme activation for organic solvents made easy, *Trends Biotechnol.*, *26*, 54-48.

[22] Rariy, R. & Klibanov, A. M. (1997) Correct protein folding in glycerol, *Proc. Natl. Acad. Sci. U.S.A.*, *94*, 13520-13523.

[23] Sirotkin, V. A., Hüttl, R. & Wolf, G., (2008). Enzyme-catalysed hydrolysis of L-amino acid esters in a low water organic solvent studied by isothermal calorimetry, *J. Therm. Anal. Calorim*, *93*, 515-520.

[24] Sirotkin, V. A. & Faizullin, D. A. (2004). Interaction enthalpies of solid human serum albumin with water-dioxane mixtures: comparison with water and organic solvent vapor sorption, *Thermochim. Acta*, *415*, 127-133.

[25] Partridge, J., Hutcheon, G. A., Moore, B. D. & Halling P. J. (1996). Exploiting hydration hysteresis for high activity of cross-linked subtilisin crystals in acetonitrile, *J. Amer. Chem. Soc.*, *118*, 12873-12877.

[26] Borisover, M. D., Sirotkin, V. A. & Solomonov, B. N. (1995). Isotherm of water sorption by human serum albumin in dioxane: Comparison with calorimetric data. *J. Phys. Org. Chem.*, *8*, 84-88.

[27] Sirotkin, V. A., Borisover, M. D. & Solomonov, B. N. (1995). Heat effects and water sorption by human serum albumin on its suspension in water-dimethyl sulfoxide mixtures. *Thermochim. Acta*, 256, 175-183.

[28] McMinn, J. H., Sowa, M. J., Charnick, S. B. & Paulaitis, M. E. (1993). The hydration of proteins in nearly anhydrous organic solvent suspensions, *Biopolymers.*, 33, 1213-1224.

[29] Kijima, T., Yamamoto, S. & Kise, H. (1996). Study of tryptophan fluorescence and catalytic activity of α-chymotrypsin in aqueous organic media, *Enz. Microb. Technol.*, 18, 2-6.

[30] Khmelnitsky, Yu., Mozhaev, V. V., Belova, A. B., Sergeeva, M. V. & Martinek, K. (1991). Denaturation capacity: a new quantitative criterion for selection of organic solvents as reaction media in biocatalysis, *Eur. J. Biochem.*, 198, 31-41.

[31] Simon, L. M., Kotorman, M., Garab, G. & Laczko, I. (2001). Structure and activity of α-chymotrypsin and trypsin in aqueous organic media, *Biochem. Biophys. Res. Comm.*, 280, 1367-1371.

[32] Uversky, V. N., Narizhneva, N. V., Lober, G. (1997). Conformational transitions provoked by organic solvents in β-lactoglobulin: Can a molten globule like intermediate be induced by the decrease in dielectric constant? *Folding & Design.* 2, 163-172.

[33] Rezaei-Ghaleh, N., Zwecktetter, M., Nemat-Gorgani, M. (2008). Amyloidogenic potential of α-chymotrypsin in different conformational states. *Biopolymers.* 91, 28-36.

[34] Goda, S., Takano, K., Yutani, K. (2000). Amyloid protofilament formation of hen egg lysozyme in highly concentrated ethanol solution. *Protein Sci.* 9, 369-375.

[35] Holley, M., Eginton, C., Brown, L. R. (2008). Characterization of amyloidogenesis of hen egg lysozyme in concentrated ethanol solution. *BBRC.* 373, 164-168.

[36] Schrier, E. E., Ingwall, R. T., Scheraga, H. A. (1965). The effect of aqueous alcohol solutions on the thermal transition of ribonuclease. *J. Phys. Chem.* 69, 298-303.

[37] Parodi, R. M., Bianchi, E., Ciferri, A. (1973). Thermodynamics of unfolding of lysozyme in aqueous alcohol solutions. *J. Biol. Chem.* 248, 4047-4051.

[38] Fujita, Y., Miyanaga, A., Noda, Y. (1979). Effect of alcohols on the thermal denaturation of lysozyme as measured by differential scanning calorimetry. *Bull. Chem. Soc. Japan.* 52, 3659-3662.

[39] Velicelebi, G., Sturtevant, J. (1979). Thermodynamics of the denaturation of lysozyme in alcohol-water mixtures. *Biochem.* 18, 1180-1186.

[40] Brandts, J. F., Hunt, L. (1967). The thermodynamics of protein denaturation. III. The denaturation of ribonuclease in water and in aqueous urea and aqueous ethanol mixtures. *J. Amer. Chem. Soc.* 89, 4826-4838.

[41] Banipal, T. S., Singh, G. (2004). Thermodynamic study of solvation of some amino acids, diglycine and lysozyme in aqueous and mixed aqueous solutions. *Thermochim. Acta.* 412, 63-83.

[42] Timasheff, S. N. (2002). Protein-solvent preferential interactions, protein hydration, and the modulation of biochemical reactions by solvent components. *Proc. Natl. Acad. Sci. USA.*, 99, 9721-9726.

[43] Arakawa, T., Kita, Y. & Timasheff, S. N. (2007). Protein precipitation and denaturation by dimethyl sulfoxide. *Biophys. Chem.*, *131*, 62-70.

[44] Gekko, K., Ohmae, E., Kameyama, K. & Takagi, T. (1998). Acetonitrile-protein interactions: Amino acid solubility and preferential solvation. *Biochim. Biophys. Acta*, *1387*, 195-205.

[45] Kovrigin, E. L. & Potekhin, S. A. (2000). On stabilizing action of protein denaturants: Acetonitrile effect on stability of lysozyme in aqueous solutions. *Biophys. Chem.*, *83*, 45-59.

[46] Shimizu, S. & Matubayasi, N. (2014). Preferential solvation: Dividing surface vs excess numbers. *J. Phys. Chem. B.*, *118*, 3922-3930.

[47] Auton, M., Bolen, D. W. & Rosgen, J. (2008). Structural thermodynamics of protein preferential solvation: Osmolyte solvation of proteins, amino acids, and peptides. *Proteins*, *73*, 802–813.

[48] Casassa, E. F. & Eisenberg, H. (1964). Thermodynamic analysis of multicomponent solutions. *Adv. Protein Chem.*, *19*, 287–395.

[49] Tanford, C. (1969). Extension of theory of linked functions to incorporate effects of protein hydration. *J. Mol. Biol.*, *39*, 539–544.

[50] Schellman, J. A. (1987). Selective binding and solvent denaturation. *Biopolymers*, *26*, 549–559.

[51] Arakawa, T., Bhat, R. & Timasheff, S. N. (1990). Why preferential hydration does not always stabilize the native structure of globular proteins. *Biochem.*, *29*, 1924-1931.

[52] Smith, P. E. (2006). Equilibrium dialysis data and the relationships between preferential interaction parameters in biological systems in terms of Kirkwood-Buff integrals. *J. Phys. Chem. B*, *110*, 2862– 2868.

[53] Parsegian, V. A., Rand, R. P. & Rau, D. C. (2000). Osmotic stress, crowding, preferential hydration, and binding: A comparison of perspectives. *Proc. Natl. Acad. Sci. U.S.A.*, *97*, 3987–3992.

[54] Kirby Hade, E. P. & Tanford, C. (1967). Isopiestic compositions as a measure of preferential interactions of macromolecules in two-component solvents. Application to proteins in concentrated aqueous cesium chloride and guanidine hydrochloride. *J. Amer. Chem. Soc.*, *89*, 5034-5040.

[55] Kamiyama, T., Liu, H. L. & Kimura, T. (2009). Preferential solvation of lysozyme by dimethyl sulfoxide in binary solutions of water and dimethyl sulfoxide. *J. Therm. Anal. Cal.*, *95*, 353-359.

[56] Reisler, E., Haik, Y. & Eisenberg, H. (1977). Bovine serum albumin in aqueous guanidine hydrochloride solutions. Preferential and absolute interactions and comparison with other systems. *Biochem.*, *16*, 197-203.

[57] Izumi, T., Yoshimura, Y. & Inoue, H. (1980). Solvation of lysozyme in water/dioxane mixtures studied in the frozen state by NMR spectroscopy. *Arch. Biophys. Biochem.*, *200*, 444-451.

[58] Fersht, A. (1999). Structure and Mechanism in Protein Science: A Guide to Enzyme Catalysis and Protein Folding; Freeman & Co: New York.

[59] Lehninger, A. L., Nelson, D. L. & Cox, M. M. (1993). *Principles of Biochemistry*; Worth: New York.

[60] Perrin, D. D., Armarego, W. L. F. & Perrin, D. R. (1980). *Purification of Laboratory Chemicals*, Oxford: Pergamon Press.

[61] Sirotkin, V. A. & Kuchierskaya, A. A. (2017). Preferential solvation/hydration of α-chymotrypsin in water-acetonitrile mixtures. *J. Phys. Chem. B.*, *121*, 4422-4430.

[62] Sirotkin, V. A. & Kuchierskaya, A. A. (2017). Lysozyme in water-acetonitrile mixtures: Preferential solvation at the inner edge of excess hydration. *J. Chem. Phys.*, *146*, 215101-8.

[63] Sirotkin, V. A. & Kuchierskaya, A. A. (2017). α-Chymotrypsin in water-ethanol mixtures: Effect of preferential interactions. *Chem. Phys. Lett.*, *689*, 156-161.

[64] Sirotkin, V. A. & Kuchierskaya, A. A. (2017). α-Chymotrypsin in water-acetone and water-dimethyl sulfoxide mixtures: Effect of preferential solvation and hydration. *Proteins: Functions, Structure and Bioinformatics*, *85*, 1808-1819.

[65] Nikolskii, B. P. (1963) *Chemist's Handbook,* Leningrad, Russia: Goskhimizdat.

[66] Prestrelsky S. J., Tedeschi N., Arakawa T., Carpenter J. F. (1993) Dehydration-induced conformational transitions in proteins and their inhibition by stabilizers. *Biophys. J,* 65, 661-671.

[67] Van de Weert M., Harvez P. I., Hennink W. E., Crommelin D. J. A. (2001) Fourier transform infrared spectrometric analysis of protein conformation: effect of sampling and stress factors. *Anal. Biochem,* 297, 160-169.

[68] Griebenow K., Klibanov A. M. (1995) Lyophilization-induced reversible changes in the secondary structure of proteins. *Proc. Natl. Acad. Sci. USA,* 92, 10969-10976.

[69] Grdadolnik, J., Marechal, Y. (2001) Bovine serum albumin observed by infrared spectrometry. I. Methodology, structural investigation, and water uptake. *Biopolymers* **62**, 40-53.

[70] Atkins, P. W. (2006). *Physical Chemistry.* 8th ed. Oxford: Oxford University Press.

[71] Prausnitz, J. M. (1969). *Molecular Thermodynamics of Fluid-Phase Equilibria.* N.J.: Prentice-Hall, Inc., Engelwood Cliffs.

[72] Belousov, V. P. & Panov, M. Y. (1994). *Thermodynamic properties of aqueous solutions of organic substances.* Boca Raton, Fla.: CRC Press.

[73] Bell, G., Janssen, A. E. M. & Halling, P. (1996). Water activity fails to predict critical hydration level for enzyme activity in polar organic solvents: Interconversion of water concentrations and activities. *Enzym. Microb. Technol.,* 20, 471-476.

Chapter 3

PREFERENTIAL HYDRATION OF α-CHYMOTRYPSIN IN ACETONITRILE: COMPARISON WITH FTIR SPECTROSCOPY

ABSTRACT

The aim of this study is to monitor the preferential hydration of the protein macromolecules at low and high water content in water-organic mixtures. Our approach is based on the analysis of the absolute values of the water/organic solvent sorption. We applied this approach to estimate the protein stabilization/destabilization due to the preferential interactions of α-chymotrypsin with water-acetonitrile mixtures. At high water content, proteins are preferentially hydrated. At the intermediate water content, the preferential interaction changed from preferential hydration to preferential binding of acetonitrile. From infrared spectra, changes in the structure of proteins were determined through an analysis of the structure of the amide I band. Acetonitrile augments the intensity of the 1626 cm^{-1} band assigned to the intermolecular β–sheet aggregates. At low water content, the proteins are in a glassy (rigid) state. The H-bond accepting acetonitrile molecules are not effective in solvating the dehydrated protein molecules alone. Therefore, the acetonitrile molecules are preferentially excluded from the protein surface, resulting in the preferential hydration.

Keywords: preferential solvation, preferential hydration, protein, water, acetonitrile

1. INTRODUCTION

Enzyme-water interactions are well-known to play a critical role in determining the function, structure, and stability of enzyme macromolecules [1-15]. Elucidation of the processes occurring upon enzyme hydration in the presence of organic solvents is essential in a wide range of biomedical and biotechnological applications. In particular, organic solvents are widely utilized as model substances to induce amyloid fibril and molten globule formation [16-19]. Numerous debilitating diseases are associated with these protein states, including type II diabetes, Parkinson's disease, and Alzheimer's disease.

Enzyme activity constitutes an intricate function of water content in organic solvents. Typical dependencies of the enzymatic activity on the water content in organic solvents can be delineated into three parts [22-35]:

- A. The first range of concentration concerns the water-rich mixtures. In this range, one can observe hydrolytic enzyme activity. However, numerous reactions that are essential to industry, such as esterification and peptide synthesis, are suppressed in aqueous solutions as a consequence of the unfavorable shift of reaction equilibria.
- B. A sharp decline in enzyme activity was found after a certain concentration threshold of the organic solvent had been reached. The physico-chemical properties of the solvent determine the position of this minimum. The enzyme structure may be disturbed by the organic solvents through weakening of the hydrophobic interactions, by altering the electrostatic

interactions of the polar protein groups and/or by direct interaction with the biocatalysts.
C. The mixtures with low water content constitute the third concentration range. The dehydrated enzymes are in a kinetically "frozen" state. Due to the reduced conformational flexibility in the water-poor organic solvents, the enzymes remain in the active conformation. There are some advantages in employing low-water organic liquids, including the high solubility of hydrophobic reagents, the catalysis of the industrially important synthetic reactions (peptide synthesis and transesterification), the suppression of undesirable side reactions caused by water, and the enhanced thermostability.

Preferential hydration/solvation is an effective method for revealing the mechanism of the protein stabilization or denaturation [36-51]. When a protein interacts with a binary water-organic solvent mixture, the three components do not equally mix. Water or organic solvent molecules exist preferentially in the protein's solvation shell. This difference between the solvation shell and bulk solvent in the solvent components has been termed preferential solvation [36-38, 40-44, 50]. Preferential solvation is a thermodynamic quantity that describes the protein surface occupancy by the water and cosolvent molecules. This is associated with the actual numbers of water/cosolvent molecules that are in contact with the protein's surface. It was also found that the protein destabilization is directly associated with the preferential binding of the denaturant molecules to specific protein groups [36-38, 40-44, 50].

Preferential hydration is the excess of water in the protein macromolecule relative to the water concentration in the overall solvent. The preferential hydration does not always stabilize the native proteins [36]. The preferentially hydrating solvent systems can be divided into two groups: (i) The first always stabilize the protein structure. The predominant interaction in the first group is the organic solvent exclusion with the protein being essentially inert. (ii) The preferential interactions in the second group are determined by the protein surface's chemical nature. This achieves a precise balance between the binding and exclusion of the cosolvent. The chemical nature of the protein surface determines the preferential binding. The exposure of additional hydrophobic protein groups can result in it being enhanced on an unfolding protein.

Preferential interactions can be detected by a variety of methods: by the isopiestic measurements of vapor pressure, by differential densitometry, by the comparison of the refractive index increment or partial specific volume prior to and after the redistribution of solvent components across a membrane that is impermeable to the biomacromolecule, and by NMR spectroscopy [36-51]. These works were performed at high water content. No attempt has yet been made, however, to investigate the preferential hydration/solvation of the protein macromolecules at low and high water content in organic solvents simultaneously.

It was previously shown that there are several protein hydration regimes in the absence of organic solvents [1-7]. The protein and water contributions to the excess thermodynamic functions (Gibbs energies, enthalpies, volumes, and entropies) were shown to markedly depend on the hydration level [6, 7]. Changes in the excess thermodynamic functions at the lowest level of water content are primarily attributable to the addition of water. At high water content, there is no significant effect on the excess quantities from water addition. In water-rich systems, changes of excess thermodynamic functions solely reflect alterations in the protein state [6, 7]. Correspondingly, the preferential hydration/solvation contributions are likely to be different depending on the composition region. These contributions may change their sign and absolute values depending on the protein state. The relationships between the preferential hydration and preferential solvation contributions may be very different at low and high water content. Therefore, understanding the mechanism of the enzyme hydration in the presence of additives (organic solvents) requires the effective experimental approaches providing the thermodynamic information about the enzyme–water and enzyme–organic solvent interactions.

The aim of our study is to monitor the preferential hydration and solvation in the entire range of water content in water-organic mixtures at 25°C. Our approach is based on the simultaneous measurements of the absolute values of the water and organic solvent sorption. The preferential solvation/hydration parameters were calculated using the water and organic solvent sorption values. The preferential solvation/hydration parameters were compared with the corresponding changes in the protein structure that transpire regarding the interaction of the protein with organic solvent and water molecules. The effect of organic solvent on the protein structure was investigated by FTIR (Fourier Transform Infrared) spectroscopy.

Advantages of our approach:

(i) The preferential interaction parameters can be determined in the entire range of water content in organic liquids.
(ii) Our approach facilitates the individual evaluation of the Gibbs energies of water, protein, and organic solvent.

Bovine pancreatic α-chymotrypsin (CT) was chosen as a model protein since it is one of the most applied and studied in enzymology and biophysical chemistry [52, 53]. Physiological role of CT is to hydrolyze peptide bonds [52, 53]. An example of a predominantly β-sheet protein is α-chymotrypsin.

The choice of acetonitrile (AN) was determined by the following reasons:

A. This solvent is widely used in nonaqueous enzymology and reverse-phase chromatography for the separation of protein and peptide mixtures [20-27].

B. Acetonitrile is an organic solvent that is water-miscible. Consequently, the effect of this low molecular substance's on the hydration and structure of α-chymotrypsin in the entire range of water content can be investigated.
C. Acetonitrile is able to form hydrogen bonds with several hydrogen donors. In contrast to water, however, it does not possess hydrogen bond donating ability.

2. EXPERIMENTAL

2.1. Materials

Bovine pancreatic α-chymotrypsin was purchased from Sigma Chemical Co. (St. Louis, MO, U.S.A., No. C 4129, protein content > 95%; essentially salt free; EC 3.4.21.1; 66 units of specific activity for N-benzoyl-L-tyrosine ethyl ester). The molecular mass of bovine pancreatic α-chymotrypsin was determined to be 25000 Da. Acetonitrile (analytical grade, with a purity of > 99%) was purified and dried according to the recommended guidelines [54]. Water used was doubly distilled. All water-organic mixtures were prepared gravimetrically using a Precisa balance (Swiss) with a precision of 0.00001 g.

2.2. Initial Protein State

The protein powder was placed on the thermostated cell and dried using a microthermoanalyzer "Setaram" MGDTD-17S (± 0.00001 g) at 25°C and 0.1 Pa, until a constant sample weight was reached. The water content of the dehydrated protein was estimated as 0.002 ± 0.001 g water/g protein, using the Karl Fischer titration method as described previously [55-58]. This value for chymotrypsin implies that at the **zero**

hydration level there are about three water molecules strongly bound to each protein molecule.

2.3. Sorption Measurements

The protein samples were prepared according to the recommended guidelines [55, 56]. The initially dried protein samples were presented to water-alcohol vapor mixtures. The water-organic vapor mixture was consecutively flowed through a thermostated saturator filled with the water-organic mixture, and a cell containing the protein sample. Protein samples (5 - 10 mg) each were flushed by water – organic vapor mixtures, until no further mass changes were detected as described previously. The sorption equilibrium was reached after 6 h at 25°C. The schematic representation of the experimental setup is given in Ref. 55. An external ethylene glycol thermostat (RC 6, Lauda, Germany) was utilized to determine the temperature with a precision of 0.1°C. The water activity (a_1) in the vapor phase was adjusted by altering the water content in the liquid water-organic mixture.

Measurements of the water sorption (A_1) were conducted by Karl Fischer titration with a Metrohm 831 KF coulometer. Alcohol content of α-chymotrypsin (A_3) was calculated as a difference between the total sorption uptake ($A_1 + A_3$) and water content (A_1). The total sorption uptake ($A_1 + A_3$) was measured by microthermoanalyzer "Setaram" MGDTD-17S.

2.4. Fourier Transform Infrared Spectroscopic Measurements

FTIR spectra were recorded with a Nicolet MAGNA 550 infrared spectrometer. FTIR spectra were measured at 25°C as described previously [55]. Co-adding 256 scans at a spectral resolution of 2 cm^{-1}

achieved each spectrum. The sample chamber was purified with dry and carbon dioxide free air.

Glassy-like protein films casted from 2% (w/v) water solution onto the CaF_2 window at room humidity and temperature achieved the infrared spectra. A schematic representation of the experimental conditions for FTIR spectroscopic measurements of the proteins' structural characteristics is given in Ref. 55. Once the window was mounted in the sample cell, the film was dehydrated through flushing air that was dried over the P_2O_5 powder. At 25°C, relative water vapor pressure over P_2O_5 did not exceed 0.01 [59].

The protein films were flushed until there was no detection of additional spectral changes in the 3450 cm^{-1} water absorbance region, as well as a smooth line without any visible shoulders was represented by a contour on this side. The spectrum of this sample was utilized as a reference spectrum for the calculation of the difference spectra.

The initially dehydrated protein was exposed to water-organic vapor mixtures. The air flowed consecutively through the thermostated glass tube accompanied by a drying agent (P_2O_5), saturator filled with a water–organic mixture, and then through the entire sample cell. The temperature of the saturator and sample cell was 25°C. The water content in the vapor phase was manipulated by altering the water concentration in the liquid water–organic mixture.

3. RESULTS AND DISCUSSION

3.1. Sorption Isotherms

Figure 1 shows the water and acetonitrile vapor sorption isotherms for α-chymotrypsin at 25°C. Three distinct effects of acetonitrile on water binding by α-chymotrypsin were identified [55]:

(i) At low water activity ($a_1 < 0.3$), water sorption is similar in the presence and absence of organic solvent. Acetonitrile does not have much effect on the interaction between tightly bound water and enzyme at low a_1.

(ii) At the intermediate water activity, at a given a_1, AN decreases water content. This behavior indicates that the suppression in the uptake of water may be the result of competition for water-binding sites on α-chymotrypsin by acetonitrile.

(iii) At $a_1 > 0.8$, AN augments the quantity of water that is bound by CT. This phenomenon was regarded as an organic solvent-assisted result on the water binding.

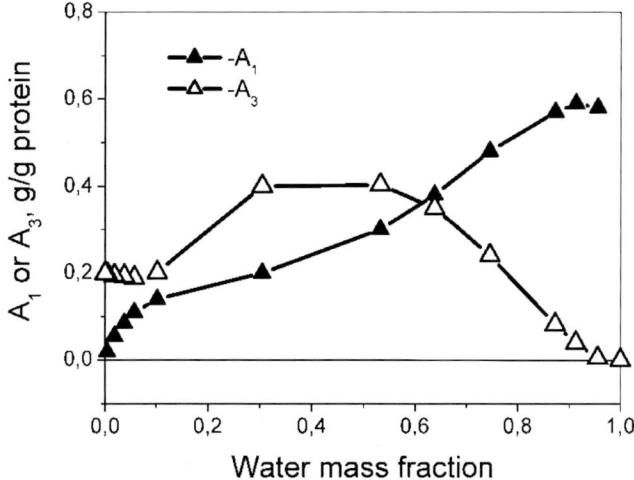

Figure 1. Water (A_1) and acetonitrile (A_3) sorption isotherms for α-chymotrypsin at 25°C.

3.2. Excess Sorption Functions

The protein solvation layer is composed of two parts: (i) nonideal (due to preferential interactions) and (ii) ideal. The nonideal effect of the solvation layer on enzyme properties (organic solvent/water sorption; residual enzyme activity) can be defined in terms of excess functions, F^E [60-62], which refers to the difference between the observed function of mixing, F^M, and the function for an ideal binary mixture, F_{id}^M.

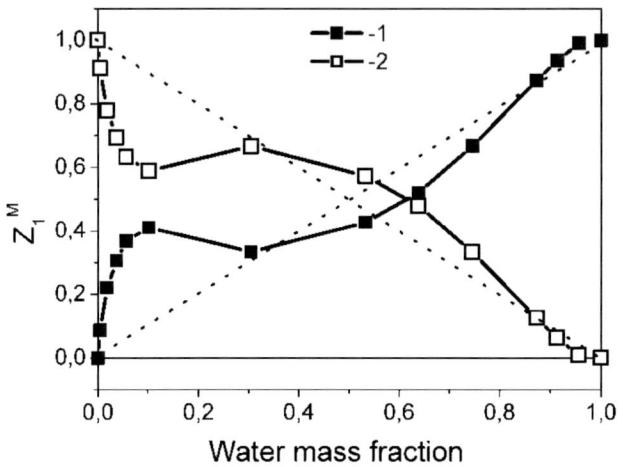

Figure 2. Water mass fraction in the solvation layer of α-chymotrypsin (Z_1^M) - (1); Acetonitrile mass fraction in the solvation layer of α-chymotrypsin (Z_3^M) – (2).

The deviations of the excess functions from zero describe the degree to which the protein solvation layer differs from the pure binary water-organic mixture as a consequence of preferential interactions between water (component 1), protein (component 2), and organic solvent (component 3).

Eq. 1 is used to calculate the F^E values:

$$F^E = F^M - F_{id}^M \tag{1}$$

The $F_{id,i}^M$ values were calculated using Eq. 2:

$$F_{id,i}^M = F_i^M(w_i = 0) + w_i[F_i^M(w_i = 1.0) - F_i^M(w_i = 0)] \tag{2}$$

where $F_i^M(w_i = 1.0)$ is the observed mixing function of protein at $w_i = 1.0$; $F_i^M(w_i = 0)$ is the observed mixing function of protein at $w_i = 0$; w_1 is the water mass fraction in the binary water-organic mixtures; and w_3 is the organic solvent mass fraction in the binary water-organic mixtures ($w_1 + w_3 = 1.0$).

The situation in which no preferential interactions exist between water, protein, and organic solvent is described by the $F_{id,i}^M$ values. In this circumstance, the water mass fraction in the ideal part of the solvation layer is the same as in the pure water-organic mixture.

The Z_1^M (water mass fraction in the solvation layer) and Z_3^M (organic solvent mass fraction in the solvation layer) values as a function of water mass fraction in acetonitrile are presented in Figure 2. Eqs. 3 and 4 are utilized to calculate the Z_1^M and Z_3^M values, respectively:

$$Z_1^M = \frac{A_1}{A_1 + A_3} \tag{3}$$

$$Z_3^M = \frac{A_3}{A_1 + A_3} \tag{4}$$

The simultaneous action of acetonitrile and water was determined by the Z_1^E and Z_3^E values (Figure 3). These excess sorption functions can be estimated using Eqs. 5 and 6:

$$Z_1^E = Z_1^M - Z_{id,1}^M \tag{5}$$

$$Z_3^E = Z_3^M - Z_{id,3}^M \tag{6}$$

where Z_1^M is the mass fraction of water in the solvation layer for the real water-organic mixture; and $Z_{id,1}^M$ is the mass fraction of water in the solvation layer for the ideal water-organic mixture. The $Z_{id,1}^M$ values were calculated using Eq. 7:

$$Z_{id,1}^M = Z_1^M(w_1 = 0) + w_1[Z_1^M(w_1 = 1.0) - Z_1^M(w_1 = 0)] \quad (7)$$

where $Z_1^M(w_1 = 1.0)$ is the water mass fraction in the solvation layer of protein at w₁=1.0; $Z_1^M(w_1 = 0)$ is the water mass fraction in the solvation shell at w₁=0; and w₁ is the mass fraction of water in organic solvent.

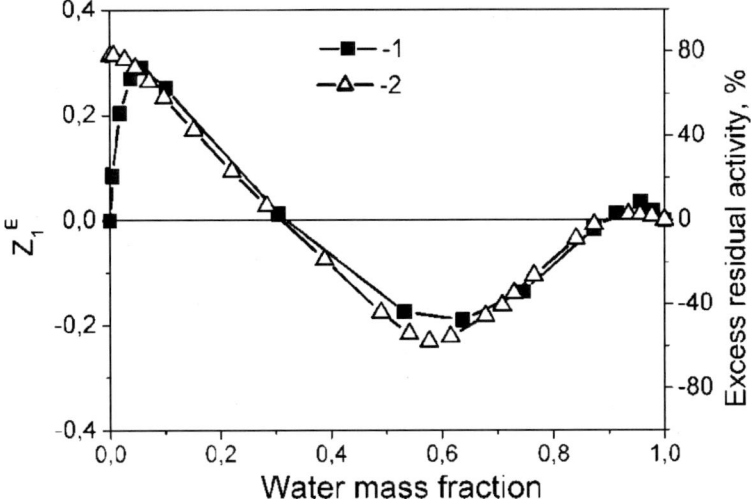

Figure 3. Excess water mass fraction in the solvation layer of α-chymotrypsin (Z_1^E) – (1); Excess residual enzyme activity in water-acetonitrile mixtures (R^E) – (2).

In addition, Z_3^M is the organic solvent mass fraction in the solvation layer for the real water-organic mixture; and $Z_{id,3}^M$ is the acetonitrile mass fraction for the ideal water-organic mixture. Eq. 8 is used to calculate the $Z_{id,3}^M$ values:

$$Z_{id,3}^M = Z_3^M(w_3 = 0) + w_3[Z_3^M(w_3 = 1.0) - Z_3^M(w_3 = 0)] \quad (8)$$

where $Z_3^M(w_3 = 0)$ is the organic mass fraction in the solvation shell of protein at w₃=1.0; $Z_3^M(w_3 = 0)$ is the organic solvent mass fraction in the solvation shell of protein at $w_3 = 0$; and w_3 is the organic solvent mass fraction in the binary water-organic mixture.

As concluded from Figure 3, a considerable decrease in the uptake of water was observed in the intermediate range of water content. The Z_1^E values are negative in this range. The most pronounced suppression was observed in the water mass fraction range from 0.5 to 0.8. On the other hand, the Z_3^E values are positive in the intermediate range. This region corresponds to the preferential solvation (negative preferential hydration) of α-chymotrypsin.

The Z_1^E values are positive at low ($w_1 = 0 - 0.2$) and high water content ($w_1 = 0.9 - 1.0$). These regions correspond to the preferential hydration of α-chymotrypsin.

3.3. Residual Enzyme Activity

Figure 4 shows typical kinetic curves for the enzymatic reaction catalyzed by α-chymotrypsin preliminary incubated in water-acetonitrile mixtures. The catalytic activity was characterized by the ratio of the extent of hydrolysis attained within 200 s with α-chymotrypsin incubated in a water-organic mixture to the same quantity measured using lysozyme incubated in pure water (Figure 4A, curve 1).

The residual activity values are presented in Figure 4B. As concluded from Figure 4B, AN affects the catalytic activity of the enzyme in a complicated way. At high water content ($w_1 \sim 0.9 - 1.0$), the residual activity values are close to 100%. At $w_1 < 0.9$, there is a sharp transition from the water-rich region to the intermediate one. The residual catalytic activity of lysozyme changes from 100 to 0% in the transition region. A minimum on the residual activity curve was observed at w_1 of ~ 0.5 in AN.

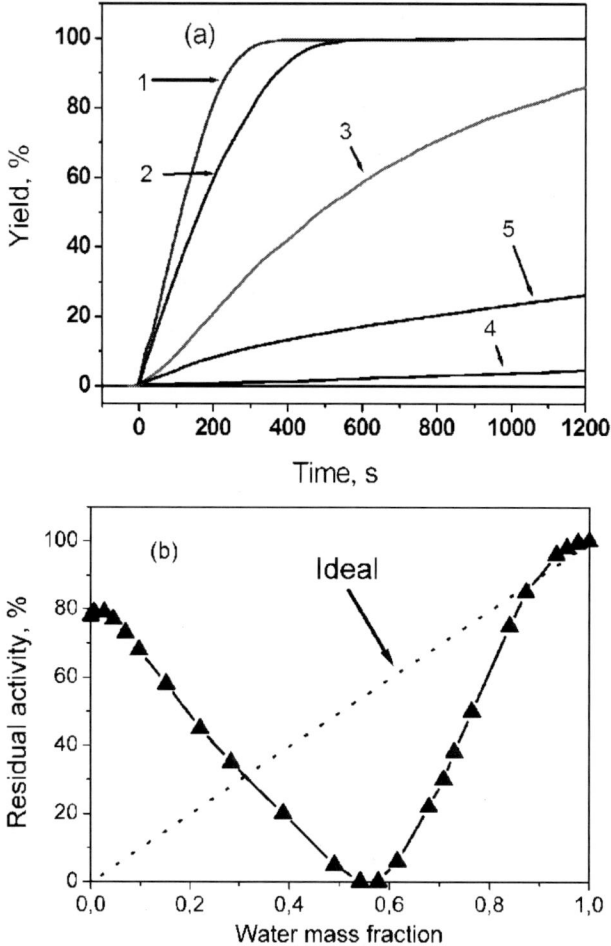

Figure 4. (a) Typical kinetic curves for the enzymatic reaction catalyzed by α-chymotrypsin previously incubated in water-acetonitrile mixtures. Water mass fraction in acetonitrile: (1) 1.0, (2) 0.005, (3) 0.85; (4) 0.49; (5) 0.28. (b) Residual activity of α-chymotrypsin in water-AN mixtures.

At $w_1 < 0.4$, the residual catalytic activity increases. At low water content in AN, the residual catalytic activity remains virtually constant, equal to ~ 80 - 85% compared with that observed after incubation in pure water.

3.4. Excess Residual Enzyme Activity

The effect of preferential interactions of the residual enzyme activity was characterized by the the R^E values (excess residual enzyme activity). The R^E values were calculated using Eq. 9:

$$R^E = R^M - R^M_{id} \tag{9}$$

where R^M is the observed residual enzyme activity; R^M_{id} is the function for an ideal binary mixture.

The R^M_{id} values were calculated using Eq. 10:

$$R^M_{id} = R^M(w_1 = 0) + w_1[R^M(w_1 = 1.0) - R^M(w_1 = 0)] \tag{10}$$

where $R^M(w_1 = 1.0)$ is the residual activity of α-chymotrypsin at $w_1 = 1.0$; $R^M(w_1 = 0)$ is the residual activity of α-chymotrypsin at $w_1 = 0$; w_1 is the water mass fraction in the binary water-organic mixtures; w_3 is the acetonitrile mass fraction in the binary water-organic mixtures ($w_1 + w_3 = 1.0$).

Figure 3 shows the dependency of the R^E values on the water mass fraction in acetonitrile. In ideal binary mixtures (mixtures of two components, W [water] and S [organic solvent]) the average W-S interactions in the solvation shell are the same as the average W-W and S-S interactions in the bulk solvent. Non-ideal mixtures are composed of particles for which the W-W, S-S, and W-S interactions are all different. As shown in Figure 3, the R^E values differ significantly from zero, indicating that the effect of the water-organic solvation layer on the residual enzyme activity is non-ideal in the entire range of water content. It is worth noting that the R^E values are consistent with the Z^E_1 values (Figure 3).

Three concentration regimes were observed:

(i) At $w_1 > 0.9$, the R^E and Z_1^E values are positive.
(ii) At the intermediate water content ($w_1 = 0.4 - 0.8$), the R^E and Z_1^E values are negative.
(iii) At low water content, the R^E and Z_1^E values are positive.

3.5. Preferential Solvation Parameters

The degree to which the solvation layer differs from the pure binary water-acetonitrile mixture due to preferential interactions between water (component 1), protein (component 2), and acetonitrile (component 3) can be characterized by the preferential interaction parameters (Eqs. 11 and 12). Eq. 11 [36-38] was utilized to determine the preferential binding of α-chymotrypsin in water-organic mixtures:

$$(\partial g_3/\partial g_2)_{T,\mu_1,\mu_3} = A_3 - \frac{w_3}{w_1} A_1 \qquad (11)$$

where A_1 is the protein hydration, expressed as gram water per gram protein; A_3 is the binding of AN, expressed as gram organic solvent per gram protein; w_1 is the water mass fraction in water-organic mixtures; and w_3 is the mass fraction of organic solvent in water-acetonitrile mixtures ($w_1 + w_3 = 1.0$).

Eq. 12 was used to estimate the preferential hydration:

$$(\partial g_1/\partial g_2)_{T,\mu_1,\mu_3} = -(\frac{w_1}{w_3})(\partial g_3/\partial g_2)_{T,\mu_1,\mu_3} \qquad (12)$$

Figures 7 and 9 show the preferential interaction parameters calculated using Eqs. 11 and 12, respectively.

Eqs. 13 and 14 were utilized to determine the Gibbs energies of the transfer of water (ΔG_1^{pref}) and acetonitrile (ΔG_3^{pref}) from water-organic mixtures to the solvation layer of α-chymotrypsin:

$$\Delta G_1^{pref} = \mu_1^E(solvation\ layer) - \mu_1^E(binary\ mixture) \quad (13)$$

$$\Delta G_3^{pref} = \mu_3^E(solvation\ layer) - \mu_3^E(binary\ mixture) \quad (14)$$

The $\mu_1^E(binary\ mixture)$ and $\mu_3^E(binary\ mixture)$ values can be calculated using Eqs. 15 and 16:

$$\mu_1^E(binary\ mixture) = RTln\gamma_1(binary\ mixture) \quad (15)$$

$$\mu_3^E(binary\ mixture) = RTln\gamma_3(binary\ mixture) \quad (16)$$

Water activity coefficients ($\gamma_1(binary\ mixture)$, the mass fraction scale; the reference state is pure water) in aqueous-organic mixtures were estimated using Eq. 17:

$$\gamma_1(binary\ mixture) = \frac{a_1}{w_1} \quad (17)$$

Organic solvent activity coefficients ($\gamma_3(binary\ mixture)$, the mass fraction scale; the reference state is pure alcohol) in aqueous-organic mixtures were calculated using Eq. 18:

$$\gamma_3(binary\ mixture) = \frac{a_3}{w_3} \quad (18)$$

Water activity (a_1) and acetonitrile activity (a_3) were taken from the published data [63] based on the vapor-liquid equilibrium.

The $\mu_1^E(solvation\ layer)$ and $\mu_3^E(solvation\ layer)$ values were calculated using Eqs. 19 and 20:

$$\mu_1^E(solvation\ layer) = RTln\gamma_1(solvation\ layer) \quad (19)$$

$$\mu_3^E(solvation\ layer) = RTln\gamma_3(solvation\ layer) \quad (20)$$

Water activity coefficients ($\gamma_1(solvation\ layer)$, the mass fraction scale) in the solvation layer were calculated using Eq. 21:

$$\gamma_1(solvation\ layer) = \frac{a_1}{Z_1^M} \qquad (21)$$

where Z_1^M is the mass fraction of water in the solvation layer;

$$Z_1^M = \frac{A_1}{A_1+A_3}.$$

Acetonitrile activity coefficients ($\gamma_3(solvation\ layer)$, the mass fraction scale) in the solvation layer were estimated using Eq. 22:

$$\gamma_3(solvation\ layer) = \frac{a_3}{Z_3^M} \qquad (22)$$

where Z_3^M is the acetonitrile mass fraction in the solvation layer;

$$Z_3^M = \frac{A_3}{A_1+A_3}.$$

Figure 5 shows the ΔG_3^{pref} and ΔG_1^{pref} values. Figures 5 and 6 demonstrate that the ΔG_3^{pref} and ΔG_1^{pref} values correlate well with the preferential interaction parameters.

The ΔG_1^{pref} value at $w_1=0$ (Figure 5) was compared with the excess Gibbs energy of water bound to the dried protein [6]. This Gibbs energy was determined from the water sorption experiments without acetonitrile. As shown in Figure 5, strong agreement between our data and previously published results was observed. This fact confirms the reliability of our calculations.

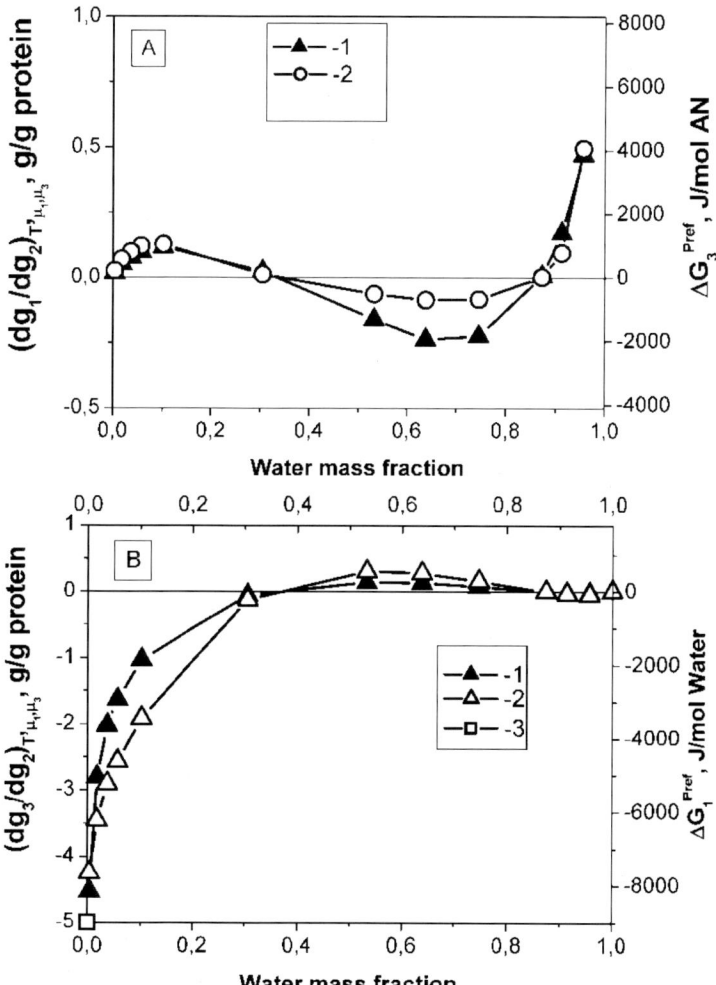

Figure 5. The preferential interaction parameters as a function of water mass fraction in acetonitrile: (A1) - $(\partial g_1/\partial g_2)_{T,\mu 1,\mu 3}$; (B1) - $(\partial g_3/\partial g_2)_{T,\mu 1,\mu 3}$. Gibbs energy of the transfer of water (B2 - ΔG_1^{pref}) and acetonitrile (A2 - ΔG_3^{pref}) from binary water-organic mixtures to the solvation shell. (B3) – Excess Gibbs energy of water bound to the dried protein [6]. Reference state is pure liquid water at 25°C.

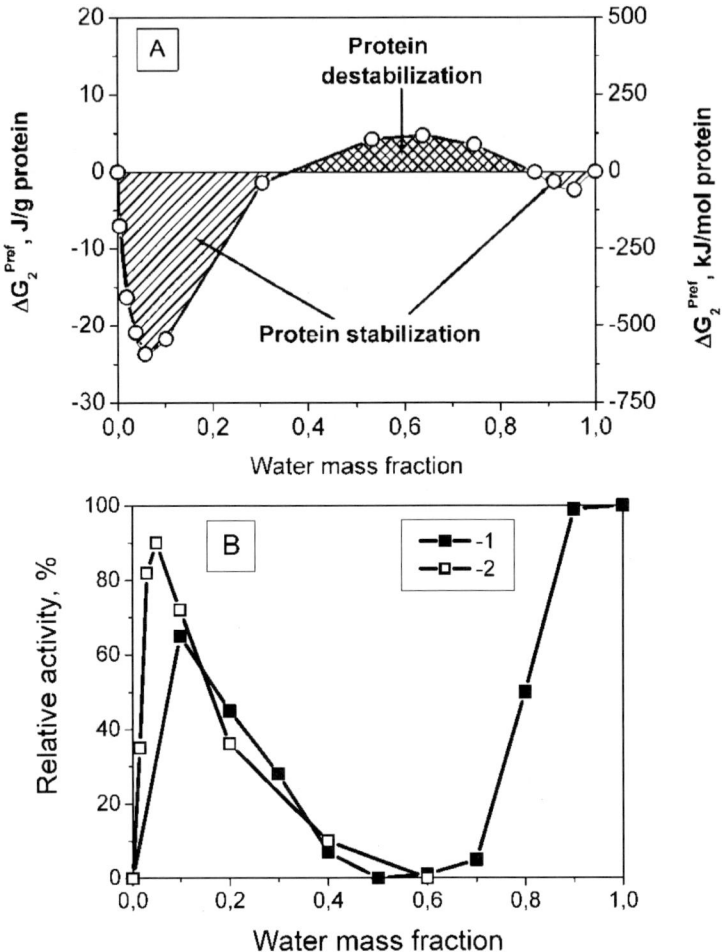

Figure 6. (A) Gibbs energy (ΔG_2^{pref}) of the transfer of CT from pure state to the water-acetonitrile mixtures. (B) Effect of water content on the reactions of N-acetyl-L-tyrosine ethyl ester catalyzed by CT in acetonitrile: (1) yield of hydrolysis [68]; (2) yield of transesterification [69].

ΔG_2^{pref} values (Figure 6) described α-chymotrypsin stabilization and destabilization due to the preferential interactions. The Gibbs-Duhem equation for ternary systems (Eq. 23) was utilized to calculate the ΔG_2^{pref} values:

$$\Delta G_2^{pref} = \frac{m_1 \Delta G_1^{pref} + m_3 \Delta G_3^{pref}}{m_2} \qquad (23)$$

$$\Delta G_2^{pref} = \mu_2^E(protein\ in\ water - alcohol\ mixtures) - \mu_2^E(pure\ protein) \qquad (24)$$

where μ_1^E, μ_2^E, and μ_3^E are the excess chemical potentials of water, protein, and acetonitrile, respectively; and m_1, m_2, and m_3 are the masses of water, protein, and organic solvent, respectively.

3.6. Infrared Spectra of α-Chymotrypsin

Infrared spectra of proteins contain some vibrational bands yielded by the amide groups of the protein backbone. Among these, the amide I band between 1700 and 1600 cm^{-1}, which is primarily a consequence of the C=O stretch of the peptide linkages, is the most sensitive to alterations in the proteins' secondary structure. The infrared spectra of α-chymotrypsin in the amide I region are presented in Figures 7 and 8. Figure 7 shows typical absorbance and second derivative spectra of α-chymotrypsin in pure water and in the dried state. The amide I contour comprises several bands at frequencies which are characteristic of particular kinds of secondary structures. It has been well-established that the frequency at the maximum absorbance of the amide I band depends strongly on the predominant type of secondary structure within the protein [64-67]. Between 1658 and 1654 cm^{-1} constitutes the maximum absorbance of the amide I band of proteins with a high content of α-helix. On the other hand, the maximum is found between 1640 and 1634 cm^{-1} with a high content of β-sheet structure.

Figure 7 shows that the band at 1638 cm^{-1} is the most dominant band component of the chymotrypsin spectra, which is attributed to the native β-sheet structure [64-67]. The band at 1685 cm^{-1} was designated for the β-sheet structure [64-67]. A minor component in the 1670 - 1665 cm^{-1} area was given to irregular secondary structures (extended chains and β-turns) [64-67].

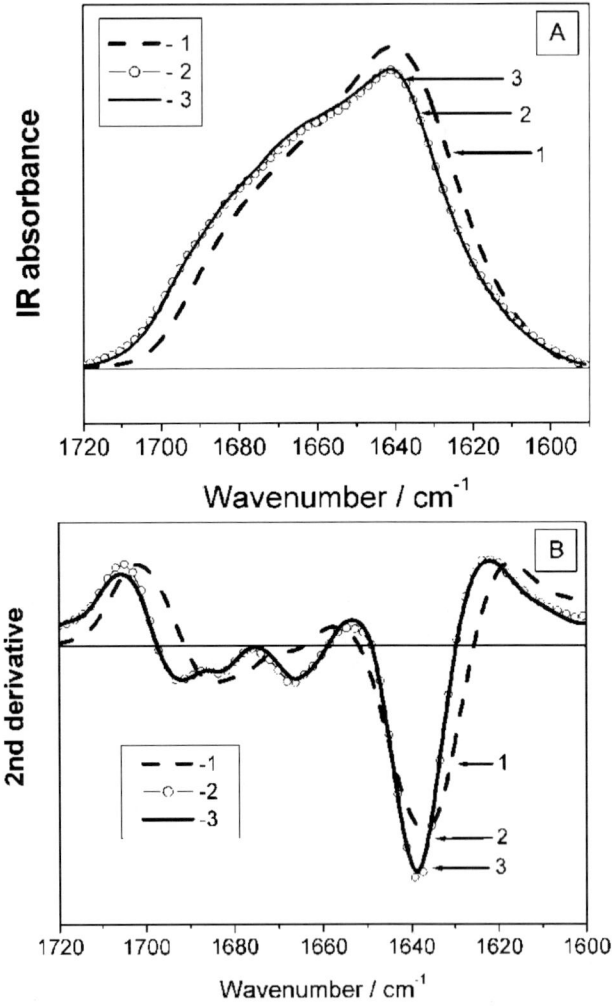

Figure 7. (A) Typical absorbance spectra of α-chymotrypsin in the amide I area at low water content: (1) CT in pure water; (2) Dried CT in the absence of AN ($A_1 < 0.002$ g/g); (3) CT in pure acetonitrile ($A_1 < 0.002$ g/g). (B) The second-derivative spectra of α-chymotrypsin within the amide I region: (1) CT in pure water; (2) Dried CT in the absence of AN ($A_1 < 0.002$ g/g); (3) CT in pure acetonitrile ($A_1 < 0.002$ g/g).

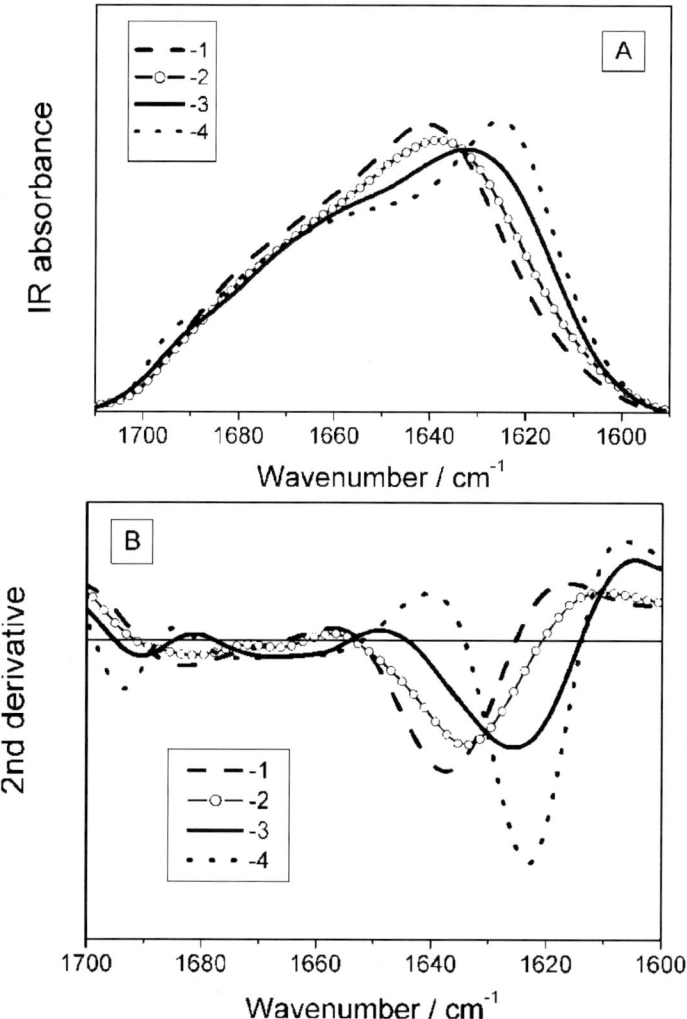

Figure 8. (A) Typical absorbance spectra of α-chymotrypsin in the amide I region at high water content in the presence of AN: (1) CT in pure water ($w_1 = 1.0$); (2) $w_1 = 0.91$; (3) $w_1 = 0.70$; (4) $w_1 = 0.63$. (B) The second-derivative spectra of α-chymotrypsin within the amide I region in the presence of AN: (1) CT in pure water ($w_1 = 1.0$); (2) $w_1 = 0.91$; (3) $w_1 = 0.70$; (4) $w_1 = 0.63$.

Figure 8 presents the absorbance and second derivative spectra of α-chymotrypsin in the presence of acetonitrile at high and intermediate water content. At high water content ($w_1 = 0.9 - 1.0$), the spectra of α-chymotrypsin are close to that for pure water. The band at 1634 - 1638 cm^{-1} constitutes the most dominant band component of the α-chymotrypsin spectra.

At the intermediate water content, the spectra of α-chymotrypsin in the presence of AN are very different from that in pure water. As the water mass fraction was decreased to 0.5 - 0.7, the amide I absorbance maxima was red-shifted to 1626 cm^{-1}. This band at 1626 cm^{-1} was assigned to the intermolecular β-sheet aggregates [64-67]. The most pronounced effect of AN was observed in the water mass fraction range of 0.5 - 0.7 (Figure 8).

3.7. Infrared Spectra of α-Chymotrypsin

Infrared spectra of proteins contain some vibrational bands yielded by the amide groups of the protein backbone. Among these, the amide I band between 1700 and 1600 cm^{-1}, which is primarily a consequence of the C=O stretch of the peptide linkages, is the most sensitive to alterations in the proteins' secondary structure. The infrared spectra of α-chymotrypsin in the amide I region are presented in Figures 7 and 8. Figure 7 shows typical absorbance and second derivative spectra of α-chymotrypsin in pure water and in the dried state. The amide I contour comprises several bands at frequencies which are characteristic of particular kinds of secondary structures. It has been well-established that the frequency at the maximum absorbance of the amide I band depends strongly on the predominant type of secondary structure within the protein [64-67]. Between 1658 and 1654 cm^{-1} constitutes the maximum absorbance of the amide I band of proteins with a high content of α-helix. On the other hand, the maximum is found between 1640 and 1634 cm^{-1} with a high content of β-sheet structure.

Figure 7 shows that the band at 1638 cm^{-1} is the most dominant band component of the chymotrypsin spectra, which is attributed to the native β-sheet structure [64-67]. The band at 1685 cm^{-1} was designated for the β-sheet structure [64-67]. A minor component in the 1670 - 1665 cm^{-1} area was given to irregular secondary structures (extended chains and β-turns) [64-67].

Figure 8 presents the absorbance and second derivative spectra of α-chymotrypsin in the presence of acetonitrile at high and intermediate water content. At high water content (w_1 = 0.9 - 1.0), the spectra of α-chymotrypsin are close to that for pure water. The band at 1634 - 1638 cm^{-1} constitutes the most dominant band component of the α-chymotrypsin spectra.

At the intermediate water content, the spectra of α-chymotrypsin in the presence of AN are very different from that in pure water. As the water mass fraction was decreased to 0.5 - 0.7, the amide I absorbance maxima was red-shifted to 1626 cm^{-1}. This band at 1626 cm^{-1} was assigned to the intermolecular β-sheet aggregates [64-67]. The most pronounced effect of AN was observed in the water mass fraction range of 0.5 - 0.7 (Figure 8).

3.8. Effect of Preferential Interactions on the Hydration and Structure of α-Chymotrypsin

The sorption, enzyme activity, and spectral experiments can be summarized as follows. Three different concentration regimes were observed:

(i) At high water content, CT is in the native (preferentially hydrated) state. The ΔG_1^{pref}, ΔG_2^{pref}, and $(\partial g_3/\partial g_2)_{T,\mu_1,\mu_3}$ values are negative. On the other hand, the ΔG_3^{pref} and $(\partial g_1/\partial g_2)_{T,\mu_1,\mu_3}$ values are positive. Hydrolytic activity with a high selectivity was observed in this range [68] (Figure 6). This indicates a lower affinity of CT for acetonitrile than water. A deficiency of AN exists near the protein surface relative to its bulk phase concentration. Our conclusion is in agreement with previously published results for the water-rich acetonitrile mixtures [38, 39]. From the experiments on the equilibrium dialysis [38] and thermal stability [39], it was found that the native protein is preferentially hydrated at room temperature.

(ii) At the intermediate water content ($w_1 = 0.4 - 0.8$), the preferential interaction changed from preferential hydration to preferential binding of AN. CT has a higher affinity for AN than for water. The acetonitrile-induced denaturation of CT is accompanied by an increase in the intensity of the 1626 cm^{-1} band given to the intermolecular β–sheet structure. The conformational changes observed may be a result of the augmented binding of AN to hydrophobic side chains, which are newly exposed on the organic solvent-induced protein denaturation. The ΔG_1^{pref}, ΔG_2^{pref} and $(\partial g_3/\partial g_2)_{T,\mu_1,\mu_3}$ values are positive. On the other hand, the ΔG_3^{pref} and $(\partial g_1/\partial g_2)_{T,\mu_1,\mu_3}$ values are negative. No noticeable enzyme activity was observed in this range (Figure 6).

(iii) At low water content, the proteins are in a glassy (rigid) state as a consequence of hydrogen bonding between polar and ionizable protein groups, and the transfer of protons [1-7]. The potential of a solvent to establish hydrogen bonds has already been demonstrated to constitute a critical factor that controls the structure and state of dehydrated proteins at room temperature [70]. Thus, H-accepting and H-donating properties of AN are anticipated to be essential for estimating the possible effects of organic molecules on the structure of the protein. When a disruption occurs with the hydrogen bond mediated protein-protein contact in the dehydrated protein, sorbate molecules (acetonitrile or water) may discern the disrupted contact's H-donating and H-accepting fragments. Water (H-acceptor and H-donor), however, is capable of solvating both H-donating and H-accepting groups. An H-accepting acetonitrile molecule is anticipated to prefer H-donating groups, whereas the remaining H-bond of the H-accepting partner will be solvated by water more efficiently. For this reason, it is anticipated that acetonitrile molecules are ineffective in solvating the dehydrated protein alone. This means that the AN molecules are preferentially excluded from the protein surface, resulting in the preferential hydration. Therefore, no noticeable acetonitrile-induced structural changes at 1626 cm^{-1} were observed at low water content (w_1 = 0 - 0.2). The ΔG_1^{pref}, ΔG_2^{pref} and $(\partial g_3/\partial g_2)_{T,\mu_1,\mu_3}$ values are negative (Figures 4 and 5). Enzymes catalyze the synthetic reactions (transesterification) at low water content [69] (Figure 6).

REFERENCES

[1] Gregory, R. B. (1995). Protein hydration and glass transition behavior. In: *Protein-Solvent Interactions* (Gregory, R.B., ed.), Marcel Dekker, New York, 191-264.

[2] Rupley, J. A. & Careri, G. (1991). Protein hydration and function. *Adv. Protein Chem.*, *41*, 37-172.

[3] Kuntz, I. D. & Kauzmann, W. (1974). Hydration of proteins and polypeptides. *Adv. Protein Chem.*, *28*, 239-345.

[4] Oleinikova, A., Smolin, N., Brovchenko, I., Geiger, A. & Winter, R. (2005). Properties of spanning water networks at protein surfaces. *J. Phys. Chem. B.*, *109*, 1988-1998.

[5] Durchschlag, H. & Zipper, P. (2001). Comparative investigations of biopolymer hydration by physicochemical and modeling techniques. *Biophys. Chem.*, *93*, 141-157.

[6] Sirotkin, V. A. & Khadiullina, A. V. (2013). Gibbs energies, enthalpies, and entropies of water and lysozyme at the inner edge of excess hydration. *J. Chem. Phys.*, *139*, 075102/1-075102/9.

[7] Sirotkin, V. A. & Khadiullina, A. V. (2014). A study of the hydration of ribonuclease A using densitometry: Effect of the protein hydrophobicity and polarity. *Chem. Phys. Lett*, *603*, 13-17.

[8] Privalov, P. L. & Crane-Robinson, C. (2017). Role of water in the formation of macromolecular structures. *Eur. Biophys. J.*, *46*, 203-224.

[9] Bull, H. B. (1944). Adsorption of water vapor by proteins. *J. Amer. Chem. Soc.*, *66*, 1499-1507.

[10] Luscher-Mattli, M. & Ruegg, M. (1982). Thermodynamic functions of biopolymer hydration. I. Their determination by vapor pressure studies, discussed in an analysis of the primary hydration process. *Biopolymers*, *21*, 403-418.

[11] Luscher-Mattli, M. & Ruegg, M. (1982). Thermodynamic functions of biopolymer hydration. II. Enthalpy-entropy compensation in hydrophilic hydration process. *Biopolymers*, *21*, 419-429.

[12] Bone, S. (1987). Time-domain reflectometry studies of water binding and structural flexibility in chymotrypsin. *Biochem. Biophys. Acta.*, *916*, 128-134.

[13] Sirotkin, V. A. & Khadiullina, A. V. (2011). Hydration of proteins: excess partial enthalpies of water and proteins. *J. Phys. Chem. B*, *115*, 15110-15118.

[14] Sirotkin, V. A., Komissarov, I. A. & Khadiullina, A. V. (2012). Hydration of proteins: excess partial volumes of water and proteins, *J. Phys. Chem. B*, *116*, 4098-4105.

[15] Sirotkin, V. A. (2005). Effect of dioxane on the structure and hydration-dehydration of α-chymotrypsin as measured by FTIR spectroscopy. *Biochim. Biophys. Acta.*, *1750*, 17-29.

[16] Uversky, V. N., Narizhneva, N. V. & Lober, G. (1997). Conformational transitions provoked by organic solvents in β-lactoglobulin: Can a molten globule like intermediate be induced by the decrease in dielectric constant? *Folding & Design.*, *2*, 163-172.

[17] Rezaei-Ghaleh, N., Zwecktetter, M. & Nemat-Gorgani, M. (2008). Amyloidogenic potential of α-chymotrypsin in different conformational states. *Biopolymers.*, *91*, 28-36.

[18] Goda, S., Takano, K. & Yutani, K. (2000). Amyloid protofilament formation of hen egg lysozyme in highly concentrated ethanol solution. *Protein Sci.*, *9*, 369-375.

[19] Holley, M., Eginton, C. & Brown, L. R. (2008). Characterization of amyloidogenesis of hen egg lysozyme in concentrated ethanol solution. *BBRC.*, *373*, 164-168.

[20] Klibanov, A. M. (2001). Improving enzymes by using them in organic solvents, *Nature*, *409*, 241-246.

[21] Carrea, G. & Riva, S. (2000). Properties and synthetic applications of enzymes in organic solvents, *Angew. Chem. Int. Ed.*, *39*, 2226-2254.

[22] Halling, P. J. (2004). What can we learn by studying enzymes in nonaqueous media? *Phil. Trans. R. Soc. Lond. B Biol. Sci.*, *359*, 1287–1297.

[23] Micaelo, N. M. & Soares C. M. (2007). Modeling hydration mechanisms of enzymes in nonpolar and polar organic solvents, *FEBS J.*, *274*, 2424–2436.

[24] Clark, D. S. (2004). Characteristics of nearly dry enzymes in organic solvents: implications for biocatalysis in the absence of water, *Phil. Trans. R. Soc. Lond. B Biol. Sci.*, *359*, 1299–1307.

[25] Serdakowski, A. L. & Dordick, J. S. (2007). Enzyme activation for organic solvents made easy, *Trends Biotechnol.*, *26*, 54-48.

[26] Rariy, R. & Klibanov, A. M. (1997) Correct protein folding in glycerol, *Proc. Natl. Acad. Sci. U.S.A.*, *94*, 13520-13523.

[27] Sirotkin, V. A., Hüttl, R. & Wolf, G. (2008). Enzyme-catalysed hydrolysis of L-amino acid esters in a low water organic solvent studied by isothermal calorimetry, *J. Therm. Anal. Calorim*, *93*, 515-520.

[28] Sirotkin, V. A. & Faizullin, D. A. (2004). Interaction enthalpies of solid human serum albumin with water-dioxane mixtures: comparison with water and organic solvent vapor sorption, *Thermochim. Acta*, *415*, 127-133.

[29] Partridge, J., Hutcheon, G. A., Moore, B. D. & Halling P. J. (1996). Exploiting hydration hysteresis for high activity of cross-linked subtilisin crystals in acetonitrile, *J. Amer. Chem. Soc.*, *118*, 12873-12877.

[30] Borisover, M. D., Sirotkin, V. A. & Solomonov, B. N. (1995). Isotherm of water sorption by human serum albumin in dioxane: Comparison with calorimetric data. *J. Phys. Org. Chem.*, *8*, 84-88.

[31] Sirotkin, V. A., Borisover, M. D. & Solomonov, B. N. (1995). Heat effects and water sorption by human serum albumin on its suspension in water-dimethyl sulfoxide mixtures. *Thermochim. Acta*, *256*, 175-183.

[32] McMinn, J. H., Sowa, M. J., Charnick, S. B. & Paulaitis, M. E. (1993). The hydration of proteins in nearly anhydrous organic solvent suspensions, *Biopolymers.*, *33*, 1213-1224.

[33] Kijima, T., Yamamoto, S. & Kise, H. (1996). Study of tryptophan fluorescence and catalytic activity of α-chymotrypsin in aqueous organic media, *Enz. Microb. Technol.*, *18*, 2-6.

[34] Khmelnitsky, Yu., Mozhaev, V. V., Belova, A. B., Sergeeva, M. V. & Martinek, K. (1991). Denaturation capacity: a new

quantitative criterion for selection of organic solvents as reaction media in biocatalysis, *Eur. J. Biochem.*, *198*, 31-41.

[35] Simon, L. M., Kotorman, M., Garab, G. & Laczko, I. (2001). Structure and activity of α-chymotrypsin and trypsin in aqueous organic media, *Biochem. Biophys. Res. Comm.*, *280*, 1367-1371.

[36] Timasheff, S. N. (2002). Protein-solvent preferential interactions, protein hydration, and the modulation of biochemical reactions by solvent components. *Proc. Natl. Acad. Sci. USA.*, *99*, 9721-9726.

[37] Arakawa, T., Kita, Y. & Timasheff, S. N. (2007). Protein precipitation and denaturation by dimethyl sulfoxide. *Biophys. Chem.*, *131*, 62-70.

[38] Gekko, K., Ohmae, E., Kameyama, K. & Takagi, T. (1998). Acetonitrile-protein interactions: Amino acid solubility and preferential solvation. *Biochim. Biophys. Acta*, *1387*, 195-205.

[39] Kovrigin, E. L. & Potekhin, S. A. (2000). On stabilizing action of protein denaturants: Acetonitrile effect on stability of lysozyme in aqueous solutions. *Biophys. Chem.*, *83*, 45-59.

[40] Shimizu, S. & Matubayasi, N. (2014). Preferential solvation: Dividing surface vs excess numbers. *J. Phys. Chem. B.*, *118*, 3922-3930.

[41] Auton, M., Bolen, D. W. & Rosgen, J. (2008). Structural thermodynamics of protein preferential solvation: Osmolyte solvation of proteins, amino acids, and peptides. *Proteins*, *73*, 802–813.

[42] Casassa, E. F. & Eisenberg, H. (1964). Thermodynamic analysis of multicomponent solutions. *Adv. Protein Chem.*, *19*, 287−395.

[43] Tanford, C. (1969). Extension of theory of linked functions to incorporate effects of protein hydration. *J. Mol. Biol.*, *39*, 539−544.

[44] Schellman, J. A. (1987). Selective binding and solvent denaturation. *Biopolymers*, *26*, 549−559.

[45] Arakawa, T., Bhat, R. & Timasheff, S. N. (1990). Why preferential hydration does not always stabilize the native structure of globular proteins. *Biochem.*, *29*, 1924-1931.

[46] Smith, P. E. (2006). Equilibrium dialysis data and the relationships between preferential interaction parameters in biological systems in terms of Kirkwood-Buff integrals. *J. Phys. Chem. B*, *110*, 2862− 2868.

[47] Parsegian, V. A., Rand, R. P. & Rau, D. C. (2000). Osmotic stress, crowding, preferential hydration, and binding: A comparison of perspectives. *Proc. Natl. Acad. Sci. U.S.A.*, *97*, 3987−3992.

[48] Kirby Hade, E. P. & Tanford, C. (1967). Isopiestic compositions as a measure of preferential interactions of macromolecules in two-component solvents. Application to proteins in concentrated aqueous cesium chloride and guanidine hydrochloride. *J. Amer. Chem. Soc.*, *89*, 5034-5040.

[49] Kamiyama, T., Liu, H. L. & Kimura, T. (2009). Preferential solvation of lysozyme by dimethyl sulfoxide in binary solutions of water and dimethyl sulfoxide. *J. Therm. Anal. Cal.*, *95*, 353-359.

[50] Reisler, E., Haik, Y. & Eisenberg, H. (1977). Bovine serum albumin in aqueous guanidine hydrochloride solutions. Preferential and absolute interactions and comparison with other systems. *Biochem.*, *16*, 197-203.

[51] Izumi, T., Yoshimura, Y. & Inoue, H. (1980). Solvation of lysozyme in water/dioxane mixtures studied in the frozen state by NMR spectroscopy. *Arch. Biophys. Biochem.*, *200*, 444-451.

[52] Fersht, A. (1999). Structure and Mechanism in Protein Science: A Guide to Enzyme Catalysis and Protein Folding; Freeman & Co: New York.

[53] Lehninger, A. L., Nelson, D. L. & Cox, M. M. (1993). Principles of Biochemistry; Worth: New York.

[54] Perrin, D. D., Armarego, W. L. F. & Perrin, D. R. (1980). *Purification of Laboratory Chemicals*, Oxford: Pergamon Press.

[55] Sirotkin, V. A. & Kuchierskaya, A. A. (2017). Preferential solvation/hydration of α-chymotrypsin in water-acetonitrile mixtures. *J. Phys. Chem. B.*, *121*, 4422-4430.

[56] Sirotkin, V. A. & Kuchierskaya, A. A. (2017). Lysozyme in water-acetonitrile mixtures: Preferential solvation at the inner edge of excess hydration. *J. Chem. Phys.*, *146*, 215101-8.

[57] Sirotkin, V. A. & Kuchierskaya, A. A. (2017). α-Chymotrypsin in water-ethanol mixtures: Effect of preferential interactions. *Chem. Phys. Lett.*, *689*, 156-161.

[58] Sirotkin, V. A. & Kuchierskaya, A. A. (2017). α-Chymotrypsin in water-acetone and water-dimethyl sulfoxide mixtures: Effect of preferential solvation and hydration. *Proteins: Functions, Structure and Bioinformatics*, *85*, 1808-1819.

[59] Nikolskii, B. P. (1963) *Chemist's Handbook*, Leningrad, Russia: Goskhimizdat.

[60] Atkins, P. W. (2006). *Physical Chemistry*. 8th ed. Oxford: Oxford University Press.

[61] Prausnitz, J. M. (1969). *Molecular Thermodynamics of Fluid-Phase Equilibria*. N.J.: Prentice-Hall, Inc., Engelwood Cliffs.

[62] Belousov, V. P. & Panov, M. Y. (1994). *Thermodynamic properties of aqueous solutions of organic substances*. Boca Raton, Fla.: CRC Press.

[63] Bell, G., Janssen, A. E. M. & Halling, P. (1996). Water activity fails to predict critical hydration level for enzyme activity in polar organic solvents: Interconversion of water concentrations and activities. *Enzym. Microb. Technol.*, *20*, 471-476.

[64] Prestrelsky, S. J., Tedeschi, N., Arakawa, T. & Carpenter, J. F. (1993). Dehydration-induced conformational transitions in proteins and their inhibition by stabilizers. *Biophys. J*, *65*, 661-671.

[65] Van de Weert, M., Harvez, P. I., Hennink, W. E. & Crommelin, D. J. A. (2001). Fourier transform infrared spectrometric analysis

of protein conformation: effect of sampling and stress factors. *Anal. Biochem*, *297*, 160-169.

[66] Griebenow, K. & Klibanov, A. M. (1995). Lyophilization-induced reversible changes in the secondary structure of proteins. *Proc. Natl. Acad. Sci. USA*, *92*, 10969-10976.

[67] Grdadolnik, J. & Marechal, Y. (2001). Bovine serum albumin observed by infrared spectrometry. I. Methodology, structural investigation, and water uptake. *Biopolymers*, *62*, 40-53.

[68] Tomiuchi, Y., Kijima, T. & Kise, H. (1993). Fluorescence Spectroscopic Study of α-Chymotrypsin as Relevant to Catalytic Activity in Aqueous-Organic Media. *Bull. Chem. Soc.*, *66*, 1176-1181.

[69] Hasegawa, M., Yamamoto, S., Kobayashi, M. & Kise, H. (2003). Catalysis of Prorease/Cyclodextrin Complexes in Organic Solvents. Effects of Reaction Conditions and Cyclodextrin Structure on Catalytic Activity of Proteases. *Enz. Microb. Tech.*, *32*, 356-361.

[70] Sirotkin, V. A., Zinatullin, A. N., Solomonov, B. N., Faizullin, D. A. & Fedotov, V. D. (2002). Interaction enthalpies of solid bovine pancreatic chymotrypsin with organic solvents: comparison with FTIR-spectroscopic data. *Thermochim. Acta*, *382*, 151-160.

Chapter 4

PREFERENTIAL SOLVATION OF LYSOZYME IN WATER-DIMETHYL SULFOXIDE MIXTURES: GIBBS ENERGIES OF WATER AND ORGANIC SOLVENT

ABSTRACT

Water/organic solvent sorption and residual enzyme activity were investigated to monitor preferential solvation and preferential hydration of protein macromolecules in the entire range of water content in organic liquids at 25°C. This approach was applied to estimate protein destabilization/stabilization due to the preferential interactions of hen-egg-white lysozyme with water-dimethyl sulfoxide (DMSO) mixtures. This approach facilitates the individual evaluation of the Gibbs energies of water, protein, and organic solvent. There are two concentration regimes. Lysozyme is preferentially hydrated at high water content. The residual enzyme activity values are close to 100%. At low water content, lysozyme has a higher affinity for DMSO than for water. The DMSO-induced irreversible inactivation of α-chymotrypsin was found in this region. Residual enzyme activity is close to zero in this concentration range.

Keywords: protein hydration, preferential solvation, enzyme activity, hen-egg-white lysozyme, dimethyl sulfoxide (DMSO)

1. INTRODUCTION

Preferential hydration/solvation offers an effective approach for elucidating the mechanism of protein destabilization/stabilization in aqueous-organic mixtures [1-16]. When a biomacromolecule interacts with a binary water-organic solvent mixture, the mixing of the three components is unequal. Organic solvent or water molecules exist preferentially in the solvation layer of the protein. This difference between the bulk solvent and solvation layer has been termed preferential solvation [1-7]. The protein denaturation is directly associated with the binding of denaturant molecules (organic solvents, salts, urea) to specific protein groups [1, 2, 6, 8].

Preferential interactions can be identified by a number of techniques: densitometry and refractometry [1, 2, 10, 11], NMR spectroscopy [16], and the isopiestic determinations of vapor pressure [13]. No attempt has yet been made, however, to simultaneously study the preferential solvation and preferential hydration of protein macromolecules at the lowest water content in organic liquids.

Water-protein interactions play an essential role in determining the functions and stability of protein molecules [17-31]. Knowledge of processes that occur upon protein hydration in the presence of organic liquids is critical in numerous biomedical and biotechnological applications, due to their utilization as biocatalysts, downstream protein processing in protein-dissolving organic solvents, and transdermal delivery of pharmaceutical proteins [32-47]. For example, there are numerous advantages in employing the enzymes in the water-poor organic liquids [32-34]:

(i) The solubility of hydrophobic substrates is higher.
(ii) The undesirable side reactions caused by water are suppressed.

(iii) Hydrolytic enzymes in nonaqueous media can catalyze the reversed hydrolytic reactions, such as transesterification, esterification, and peptide synthesis.

Therefore, elucidating the precise mechanism of protein hydration in the presence of organic solvents necessitates effective experimental methods that reveal biothermodynamic information regarding protein–organic and protein–water interactions.

The main focus of our study is to monitor the preferential hydration and preferential solvation of the protein macromolecules at low and high water content in organic liquids at 25°C. Our approach is based on an analysis of residual enzyme activity and water/organic solvent sorption. Facilitation of individual evaluation of the Gibbs energies of water, organic solvent, and protein constitutes one of the most important advantages of our approach.

Hen egg-white lysozyme was used as a model protein. This protein is one of the most applied and studied in biophysical and biotechnological investigations [48, 49]. Lysozyme is a small monomeric protein of 129 amino acid residues. The physiological role of lysozyme is to hydrolyze polysaccharide chains [48, 49].

The choice of DMSO was determined for the following reasons:

A. DMSO is a water-miscible organic solvent. Therefore, it is possible to study the effect of this low molecular substance on the hydration and functions of α-chymotrypsin in the entire range of water content.
B. DMSO is capable of forming hydrogen bonds with various hydrogen donors. In contrast to water, however, it has no hydrogen bond donating ability.

DMSO has a strong hydrogen bond accepting ability with respect to water. The enthalpy of the specific interaction (hydrogen bonding) of water with DMSO, $\Delta H_{int}^{H_2O/S}(spec.)$, is –33.1 kJ/mol [50]. For comparison, acetone is a moderate-strength hydrogen bond accepting solvent. The enthalpy of the specific interaction of water with acetone is -20.5 kJ/mol [50]. The enthalpy of the specific interaction of water with benzene equals -1.5 kJ/mol. This means that DMSO may be considered as an informative tool for analyzing the effect of the hydrogen bond accepting ability on protein-water systems.

2. EXPERIMENTAL

2.1. Materials

Hen egg-white lysozyme (EC 3.2.1.17; crystallized three times, dialyzed, and lyophilized and dried *Micrococcus lysodeikticus* cells were purchased from Sigma Chemical Co. (St. Louis, MO, U.S.A.). The molecular weight of the protein was taken as 14300 Da. DMSO (analytical grade, with a purity of >99%) was purified and dried according to the recommended guidelines [51]. Water used was doubly distilled. All water-organic mixtures were prepared gravimetrically using a Precisa balance (Swiss) with a precision of 0.00001 g.

2.2. Initial Protein State

The lysozyme powder was placed on the thermostated cell and dried using a microthermoanalyzer "Setaram" MGDTD-17S (±0.00001 g) at 25°C and 0.1 Pa, until a constant sample weight was reached. The water content of the dehydrated protein was estimated as 0.002 ± 0.001 g waterg^{-1} protein, using the Karl Fischer titration method as described previously [52]. This value for lysozyme implies that at the *zero hydration level* there are about two water molecules strongly bound to each protein molecule.

2.3. Organic Solvent and Water Sorption Measurements

The protein samples were prepared according to the recommended guidelines [53-56]. The protein samples were presented to water-organic vapor mixtures. The water-organic vapor mixture was consecutively flowed through a thermostated saturator filled with the water-organic mixture, and a cell containing the protein sample. Protein samples (5-10 mg) each were flushed by water – organic vapour mixtures, until no further mass changes were detected as described previously [53]. The sorption equilibrium was reached after 6 h at 25°C. The schematic representation of the experimental setup is given in Ref. 53. An external ethylene glycol thermostat (RC 6, Lauda, Germany) was utilized to determine the temperature with a precision of 0.1°C. The water activity (a_1) in the vapor phase was adjusted by altering the water content in the liquid water-organic mixture.

Measurements of the protein-bound water (A_1) were conducted by Karl Fischer titration with a Metrohm 831 KF coulometer. This method has been routinely used to measure the water content of protein powders. DMSO content of lysozyme (A_3) was calculated as a difference between the total sorption uptake (A_1+A_3) and water content (A_1). The total sorption uptake ($A_1 + A_3$) was measured by microthermoanalyzer "Setaram" MGDTD-17S.

2.4. Residual Enzyme Activity

Residual enzyme activity was determined by measuring the enzyme activity after storage in water-organic mixtures according to the recommended guidelines [56]. The lysozyme activity was determined as follows. The dried lysozyme was immersed in an aqueous-organic mixture of required composition and was incubated at 25°C for 3 h. This time period exceeded the time corresponding to the completion of the calorimetric heat effect accompanying the interaction of the dried proteins with pure organic solvents and water-organic mixtures [52,57]. The concentration of lysozyme in the water-organic mixtures was 1 mg/ml. Adding 100-μl aliquots of the lysozyme solution in the water-rich ethanol/EG (or the lysozyme suspension in the water-poor ethanol/EG) to the aqueous solution of the substrate (*Micrococcus lysodeikticus* cells (2.9 ml, 0.3 mg/ml) in 0.1 M potassium phosphate pH 7.0), we initiated the enzymatic reaction. Change in the absorbance at 450 nm was detected using a Perkin-Elmer Lambda 35 double-beam scanning spectrophotometer. The reaction was followed for 300-1500 s. Each kinetic curve was reproduced not less than three times.

2.5. Densitometry

Density measurements of pure liquids and mixtures were performed at atmospheric pressure and 25°C by means of vibrating-tube densimeter (Anton Paar, Austria, DMA 5000M, precision $\pm 1 \times 10^{-6}$ g cm^{-3}). Before each series of measurements the densimeter was calibrated with distilled water and air. All water-organic mixtures were prepared gravimetrically using a Precisa balance (Swiss) with a precision of 0.00001 g.

3. RESULTS AND DISCUSSION

3.1. Sorption Isotherms and Excess Sorption Functions

Figures 1 and 2 show the water (A_1) and DMSO (A_3) vapor sorption isotherms for lysozyme at 25°C. The sorption isotherms depend markedly on the water content in DMSO.

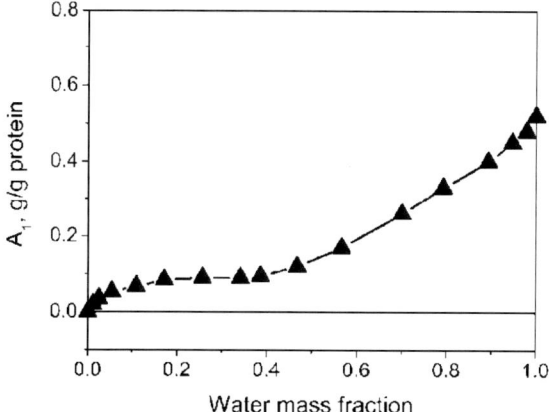

Figure 1. Water (A_1) sorption isotherm for lysozyme at 25°C. The standard errors of estimation of the water sorption were 0.001 - 0.002 g/g. Each experiment was performed 3 - 4 times.

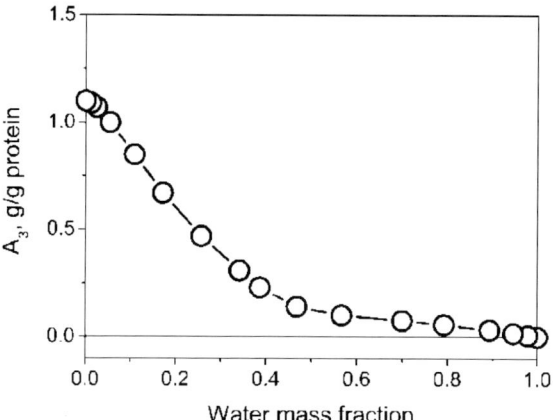

Figure 2. DMSO (A_3) sorption isotherm for lysozyme at 25°C. The standard errors of estimation of the DMSO sorption were 0.001 - 0.002 g/g. Each experiment was performed 3 - 4 times.

The protein solvation shell comprises two parts: (i) nonideal (due to preferential solvation/hydration) and (ii) ideal. The nonideal effect of the solvation shell on enzyme properties (residual enzyme activity, organic solvent and water sorption) can be articulated in terms of excess functions, F^E, [58-60] which refers to the difference between the observed function of mixing, F^M, and the function for an ideal binary mixture, F_{id}^M.

The deviations of the excess functions from zero describes the extent to which the protein solvation shell differs from the pure binary water-organic system as a consequence of preferential interactions occurring between water (component 1), protein (component 2), and organic solvent (component 3).

Eq. 1 is used to calculate the F^E values:

$$F^E = F^M - F_{id}^M \tag{1}$$

The $F_{id,i}^M$ values can be calculated using Eq. 2:

$$F_{id,i}^M = F_i^M(w_i = 0) + w_i[F_i^M(w_i = 1.0) - F_i^M(w_i = 0)] \tag{2}$$

where $F_i^M(w_i = 1.0)$ is the observed mixing function of lysozyme at $w_i = 1.0$; $F_i^M(w_i = 0)$ is the observed mixing function of lysozyme at $w_i = 0$; w_1 is the water mass fraction in the binary water-organic mixtures; and w_3 is the organic solvent mass fraction in the binary water-organic mixtures ($w_1 + w_3 = 1.0$).

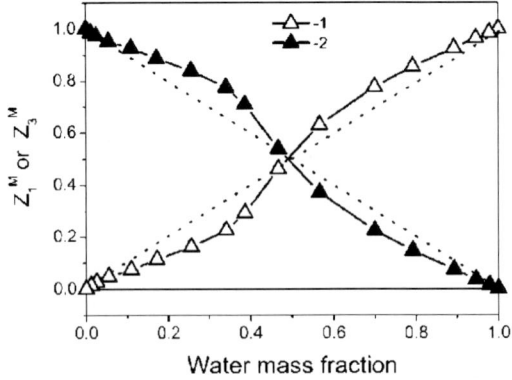

Figure 3. Water mass fraction in the solvation layer of lysozyme (Z_1^M). 2 - DMSO mass fraction in the solvation layer of lysozyme (Z_3^M).

The situation in which no preferential interactions exist between water, protein, and organic solvent is described by the $F_{id,i}^M$ values. In this circumstance, the water mass fraction in the ideal part of the solvation shell is the same as in the pure water-organic mixture.

The Z_1^M (water mass fraction in the solvation shell) and Z_3^M (organic solvent mass fraction in the solvation shell) values as a function of water mass fraction in DMSO are shown in Figure 4. Eqs. 3 and 4 are used to calculate the Z_1^M and Z_3^M values, respectively:

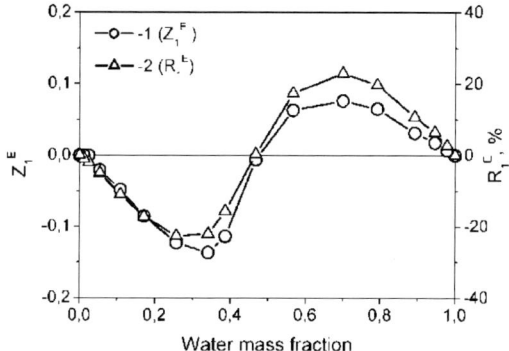

Figure 4. 1 - Excess water mass fraction in the solvation layer of lysozyme (Z_1^E). 2 - Excess residual activity of lysozyme in water-DMSO mixtures (R_1^E).

$$Z_1^M = \frac{A_1}{A_1+A_3} \qquad (3)$$

$$Z_3^M = \frac{A_3}{A_1+A_3} \qquad (4)$$

The simultaneous action of DMSO and water was characterized by the Z_1^E and Z_3^E values (Figure 5). These excess sorption functions were calculated using Eqs. 5 and 6:

$$Z_1^E = Z_1^M - Z_{id,1}^M \qquad (5)$$

$$Z_3^E = Z_3^M - Z_{id,3}^M \qquad (6)$$

where Z_1^M is the mass fraction of water in the solvation layer for the real water-organic mixture; and $Z_{id,1}^M$ is the mass fraction of water in the solvation layer for the ideal water-organic mixture. The $Z_{id,1}^M$ values were calculated using Eq. 7:

$$Z_{id,1}^M = Z_1^M(w_1 = 0) + w_1[Z_1^M(w_1 = 1.0) - Z_1^M(w_1 = 0)] \quad (7)$$

where $Z_1^M(w_1 = 1.0)$ is the water mass fraction in the solvation shell of lysozyme at $w_1 = 1.0$; $Z_1^M(w_1 = 0)$ is the water mass fraction in the solvation shell at $w_1 = 0$; and w_1 is the mass fraction of water in organic solvent.

In addition, Z_3^M is the organic solvent mass fraction in the solvation shell for the real water-organic mixture; and $Z_{id,3}^M$ is the organic solvent mass fraction for the ideal water-organic mixture. Eq. 8 is used to calculate the $Z_{id,3}^M$ values:

$$Z_{id,3}^M = Z_3^M(w_3 = 0) + w_3[Z_3^M(w_3 = 1.0) - Z_3^M(w_3 = 0)] \quad (8)$$

where $Z_3^M(w_3 = 0)$ is the organic solvent mass fraction in the solvation shell of lysozyme at $w_3 = 1.0$; $Z_3^M(w_3 = 0)$ is the organic solvent mass fraction in the solvation shell of lysozyme at $w_3 = 0$; and w_3 is the organic solvent mass fraction in the binary water-organic mixture.

Figure 4 shows that the Z_1^E values are positive at high ($w_1 = 0.5$-1.0) and low ($w_1 = 0$-0.1) water content. At intermediate water content ($w_1 = 0.1$-0.5), the Z_1^E values are negative.

3.2. Residual Enzyme Activity

Figures 5 and 6 show typical kinetic curves for the enzymatic reaction catalyzed by lysozyme preliminarily incubated in water-organic mixtures. The catalytic activity was characterized by the ratio of the extent of hydrolysis attained within 300 s with lysozyme incubated in a water-organic mixture to the same quantity measured using lysozyme incubated in pure water (Figure 5, curve 1).

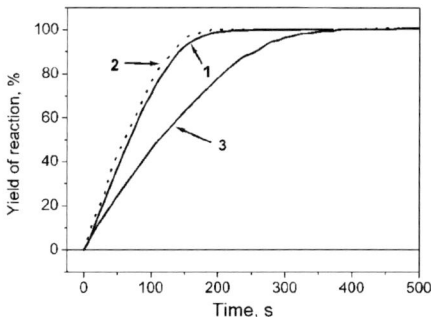

Figure 5. Typical kinetic curves for the enzymatic reaction catalyzed by lysozyme previously incubated in water-DMSO mixtures. Water mass fraction in DMSO: (1) 1.0; (2) 0.95; (3) 0.57.

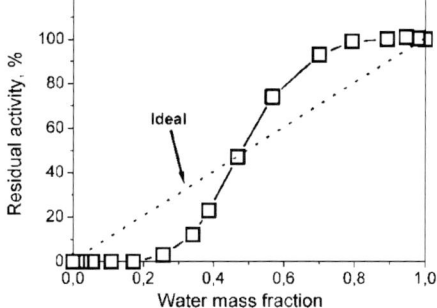

Figure 6. Residual activity of lysozyme in water-DMSO mixtures. All values are the averages of three measurements. Experimental errors were 1 - 1.5%.

The residual activity values are presented in Figure 6. As concluded from Figure 6, DMSO affects the catalytic activity of lysozyme in a complicated manner. At high water content ($w_1 \sim 0.6$-1.0), the residual activity values are close to 100% or higher (Figure 6). The residual catalytic activity of lysozyme changes from 100 to 0% in the transition region (w_1~0.3-0.6). At the lowest water content, the residual catalytic activity is close to zero.

3.3. Excess Residual Enzyme Activity

The effect of the excess hydration (Z_1^E) of the residual enzyme activity was characterized by the R_1^E values (excess residual enzyme activity) (Figure 4). The R_1^E values were calculated using Eq. 9:

$$R_1^E = R^M - R_{id,1}^M \qquad (9)$$

where R^M is the observed residual enzyme activity; and $R_{id,1}^M$ is the function for an ideal binary mixture.

The R_{id}^M values were calculated using Eq. 10:

$$R_{id,1}^M = R^M(w_1 = 0) + w_1[R^M(w_1 = 1.0) - R^M(w_1 = 0)] \qquad (10)$$

where $R^M(w_1 = 1.0)$ is the residual activity of lysozyme at $w_1 = 1.0$; $R^M(w_1 = 0)$ is the residual activity of lysozyme at $w_1 = 0$; w_1 is the water mass fraction in the binary water-organic mixtures; and w_3 is the organic solvent mass fraction in the binary water-organic mixtures ($w_1 + w_3 = 1.0$).

Figure 4 presents the dependence of the R_1^E values on the water mass fraction in DMSO. In ideal binary mixtures (mixtures of two components, W [water] and S [organic solvent]) the average W-S interactions in the solvation shell are the same as the average W-W and S-S interactions in the bulk solvent. Nonideal mixtures are composed of particles for which the W-W, S-S, and W-S interactions are all different. As shown in Figure 4, the R_1^E values differ significantly from zero, indicating that the effect of the water-organic solvation layer on the residual enzyme activity is nonideal in the entire range of water content. It is interesting to note that the R_1^E values are consistent with the Z_1^E values (Figure 4).

Two concentration regimes were observed for water-DMSO mixtures:

(i) At high and intermediate water content (Figure 4), the R_1^E and Z_1^E values are positive.
(ii) At the lowest water content, the R_1^E and Z_1^E values are negative. The DMSO-induced irreversible inactivation of lysozyme was found in this region.

3.4. Preferential Interaction Parameters

The extent to which the solvation shell differs from the pure binary water-organic system due to preferential interactions between water (component 1), protein (component 2), and organic solvent (component 3) are described by the preferential interaction parameters (Eqs. 11 and 12). Eq. 11 [1-3] is utilized to characterize the preferential binding of lysozyme in water-organic mixtures:

$$(\partial g_3/\partial g_2)_{T,\mu_1,\mu_3} = A_3 - \frac{w_3}{w_1} A_1 \qquad (11)$$

where A_1 is the protein hydration, expressed as gram water per gram protein; A_3 is the binding of organic solvent, expressed as gram organic solvent per gram protein; w_1 is the water mass fraction in water-organic mixtures; and w_3 is the mass fraction of organic solvent in water-organic mixtures ($w_1 + w_3 = 1.0$).

Eq. 12 is used to estimate the preferential hydration:

$$(\partial g_1/\partial g_2)_{T,\mu_1,\mu_3} = -\left(\frac{w_1}{w_3}\right)(\partial g_3/\partial g_2)_{T,\mu_1,\mu_3} \qquad (12)$$

Figures 7-10 show the preferential interaction parameters calculated using Eqs. 11 and 12, respectively. It is worth noting that our $(\partial g_3/\partial g_2)_{T,\mu_1,\mu_3}$ values (Figure 9) agree well with previously published data for lysozyme dissolved in water-DMSO mixtures determined by differential refractometry [2].

Eqs. 13 and 14 were utilized to determine the Gibbs energies of the transfer of water (ΔG_1^{pref}) and DMSO (ΔG_3^{pref}) from aqueous-organic mixtures to the solvation shell of lysozyme:

$$\Delta G_1^{pref} = \mu_1^E(solvation\ shell) - \mu_1^E(binary\ mixture) \quad (13)$$

$$\Delta G_3^{pref} = \mu_3^E(solvation\ shell) - \mu_3^E(binary\ mixture) \quad (14)$$

The $\mu_1^E(binary\ mixture)$ and $\mu_3^E(binary\ mixture)$ values were calculated using Eqs. 15 and 16:

$$\mu_1^E(binary\ mixture) = RT ln\gamma_1(binary\ mixture) \quad (15)$$

$$\mu_3^E(binary\ mixture) = RT ln\gamma_3(binary\ mixture) \quad (16)$$

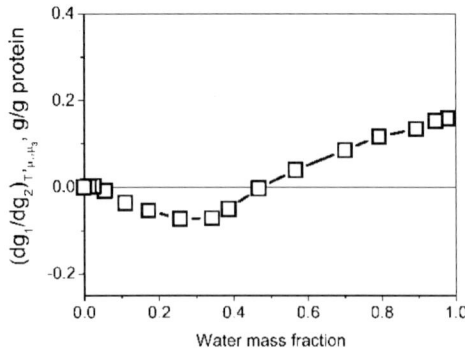

Figure 7. The preferential hydration parameter as a function of water mass fraction in DMSO ($[\partial g_1/\partial g_2]_{T,\mu_1,\mu_3}$).

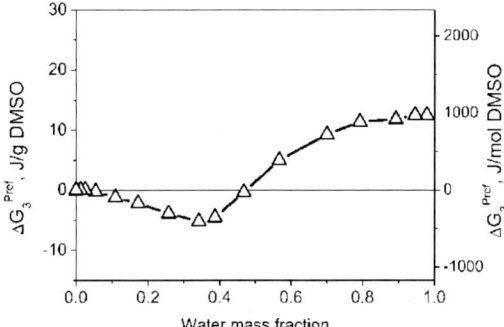

Figure 8. Gibbs energy of the transfer of DMSO (ΔG_3^{pref}) from binary water-organic mixtures to the solvation shell of lysozyme.

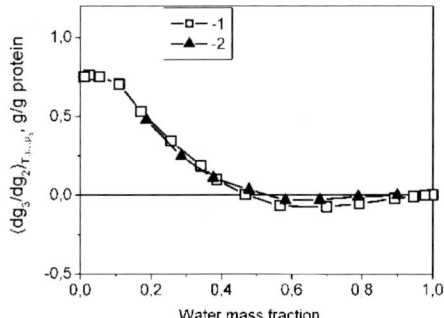

Figure 9. The preferential solvation parameter as a function of water mass fraction in DMSO ($[\partial g_3/\partial g_2]_{T,\mu_1,\mu_3}$): (1) – This work. (2) - Lysozyme. Adapted data from. Ref. 2.

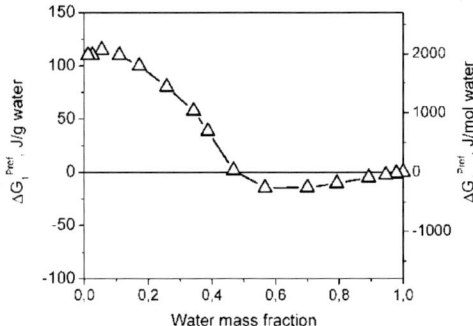

Figure 10. Gibbs energy of the transfer of water (ΔG_1^{pref}) from binary water-organic mixtures to the solvation shell of lysozyme.

Water activity coefficients ($\gamma_1(binary\ mixture)$, the mass fraction scale; the reference state is pure water) in aqueous-organic mixtures were estimated using Eq. 17:

$$\gamma_1(binary\ mixture) = \frac{a_1}{w_1} \qquad (17)$$

Organic solvent activity coefficients ($\gamma_3(binary\ mixture)$, the mass fraction scale; the reference state is pure DMSO) in aqueous-organic mixtures were calculated using Eq. 18:

$$\gamma_3(binary\ mixture) = \frac{a_3}{w_3} \qquad (18)$$

Water activity (a_1) and organic solvent activity (a_3) were taken from the published data [61] based on the vapor-liquid equilibrium.

The $\mu_1^E(solvation\ shell)$ and $\mu_3^E(solvation\ shell)$ values were estimated using Eqs. 19 and 20:

$$\mu_1^E(solvation\ shell) = RT\ln\gamma_1(solvation\ shell) \qquad (19)$$

$$\mu_3^E(solvation\ shell) = RT\ln\gamma_3(solvation\ shell) \qquad (20)$$

Water activity coefficients ($\gamma_1(solvation\ shell)$, the mass fraction scale) in the solvation shell were calculated using Eq. 21:

$$\gamma_1(solvation\ shell) = \frac{a_1}{Z_1^M} \qquad (21)$$

where Z_1^M is the mass fraction of water in the solvation shell; $Z_1^M = \frac{A_1}{A_1+A_3}$.

Organic solvent activity coefficients ($\gamma_3(solvation\ shell)$, the mass fraction scale) in the solvation shell were calculated using Eq. 22:

$$\gamma_3(solvation\ shell) = \frac{a_3}{Z_3^M} \tag{22}$$

where Z_3^M is the organic solvent mass fraction in the solvation shell; $Z_3^M = \frac{A_3}{A_1+A_3}$.

Figures 7-10 present the ΔG_1^{pref} and ΔG_3^{pref} values. Figures 7-10 show that the ΔG_1^{pref} and ΔG_3^{pref} values correlate well with the preferential interaction parameters.

The ΔG_2^{pref} values (Figure 11) characterized lysozyme destabilization/stabilization due to the preferential interactions. The Gibbs-Duhem equation for ternary systems (Eq. 23) was used to calculate the ΔG_2^{pref} values:

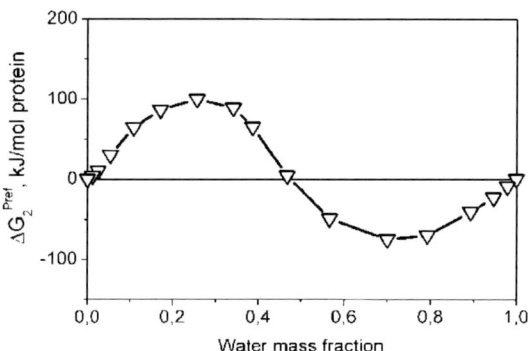

Figure 11. Gibbs energy (ΔG_2^{pref}) of the transfer of lysozyme from pure state to the water-DMSO mixtures.

$$\Delta G_2^{pref} = \frac{m_1 \Delta G_1^{pref} + m_3 \Delta G_3^{pref}}{m_2} \tag{23}$$

$$\Delta G_2^{pref} = \mu_2^E(protein\ in\ water - organic\ mixtures) - \mu_2^E(pure\ protein) \tag{24}$$

where μ_1^E, μ_2^E, and μ_3^E are the excess chemical potentials of water, protein, and organic solvent, respectively; and m_1, m_2, and m_3 are the masses of water, protein, and organic solvent, respectively.

3.5. Preferentilal Solvation in Binary Water-Organic Mixtures

Preferential solvation (hydration) in binary water-DMSO mixtures was defined as follows. The absolute values of the solvation (hydration) excess ($\Gamma_{A(B)}^k$) are estimated using Eqs. 25 and 26:

$$\Gamma_{1(2)}^k = -\left(\frac{d\ln[\gamma_1(Pref.hydration)]}{d\ln(a_1)}\right)_{T,P,a_2} \quad (25)$$

$$\Gamma_{2(1)}^k = -\left(\frac{d\ln[\gamma_2(Pref.solvation)]}{d\ln(a_2)}\right)_{T,P,a_1} \quad (26)$$

where $\Gamma_{A(B)}^k$ is the solvation excess of the A particles over the B component in the surrounding of k particles [62]; ($\gamma_1(binary\ mixture)$ and $\gamma_2(binary\ mixture)$ are the activity coefficients in the unperturbed binary mixture; $\gamma_1(solvation\ layer)$ and $\gamma_2(solvation\ layer)$ are the activity coefficients in the solvation layer; a_1 and a_2 are the thermodynamic activities. Water activity (a_1) and organic solvent activity (a_2) were taken from the published data [61] based on the vapor-liquid equilibrium.

The solvation excess ($\Gamma_{A(B)}^k$) is the average number of the A particles by which the solvation shell of the k particles is richer as compared to the B molecules. If in the solvent surrounding a k particle the ratio of the A and B particle numbers is equal to their ratio in the unperturbed bulk solution, then $\Gamma_{A(B)}^k = 0$. No preferential solvation (hydration) occurs in this case.

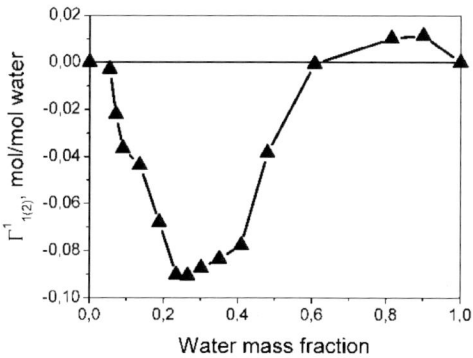

Figure 12. The hydration excess for water-DMSO mixtures at 25°C.

The hydration excess values ($\Gamma^1_{1(2)}$) for the water-dimethyl sulfoxide mixtures are given in Figure 12. Two composition regimes were observed in water-DMSO mixtures. As concluded from Figure 12, the hydration excess values are positive at high water content ($w_1 = 0.5$-1.0). This means that the solvation shell of water molecules is richer by the water molecules as compared to the unperturbed bulk solvent.

At low water content, however, the $\Gamma^1_{1(2)}$ values are negative. This fact indicates that the solvation shell of water molecules is richer by the DMSO molecules as compared to the unperturbed bulk solvent.

Thermodynamic functions of the preferential interactions in binary water-organic mixtures were determined using Eqs. 27 and 28. These equations were utilized to determine the Gibbs energies of the transfer of water [$\Delta G^{pref}_1 = RT\ln[\gamma_1(Pref.hydration)]$] and component 2 [$\Delta G^{pref}_2 = RT\ln[\gamma_2(Pref.solvation)]$] from the unperturbed binary mixture to the solvation layer of water or component 2:

$$\Delta G^{pref}_1 = \mu^E_1(solvation\ layer) - \mu^E_1(binary\ mixture) \qquad (27)$$

$$\Delta G^{pref}_2 = \mu^E_2(solvation\ layer) - \mu^E_2(binary\ mixture) \qquad (28)$$

The $\mu_1^E(binary\ mixture)$ and $\mu_2^E(binary\ mixture)$ values can be calculated using Eqs. 29 and 30:

$$\mu_1^E(binary\ mixture) = RT\ln\gamma_1(binary\ mixture) \qquad (29)$$

$$\mu_2^E(binary\ mixture) = RT\ln\gamma_2(binary\ mixture) \qquad (30)$$

Water activity coefficients ($\gamma_1(binary\ mixture)$, the mass fraction scale; the reference state is pure water) in aqueous-organic mixtures were estimated using Eq. 31:

$$\gamma_1(binary\ mixture) = \frac{a_1}{w_1} \qquad (31)$$

Organic solvent activity coefficients ($\gamma_2(binary\ mixture)$, the mass fraction scale; the reference state is pure organic solvent) in aqueous-organic mixtures were calculated using Eq. 32:

$$\gamma_2(binary\ mixture) = \frac{a_2}{w_2} \qquad (32)$$

The $\mu_1^E(solvation\ layer)$ and $\mu_2^E(solvation\ layer)$ values were calculated using Eqs. 33 and 34:

$$\mu_1^E(solvation\ layer) = RT\ln\gamma_1(solvation\ layer) \qquad (33)$$

$$\mu_2^E(solvation\ layer) = RT\ln\gamma_2(solvation\ layer) \qquad (34)$$

Water activity coefficients ($\gamma_1(solvation\ layer)$, the mass fraction scale) in the solvation layer were calculated using Eq. 35:

$$\gamma_1(solvation\ layer) = \frac{a_1}{Z_1^M} \qquad (35)$$

where Z_1^M is the mass fraction of water in the solvation layer; ρ_1^0 – the density of pure water:

$$Z_1^M = \frac{\rho_1}{\rho_1^0} w_1.$$

Organic solvent activity coefficients ($\gamma_2(solvation\ layer)$, the mass fraction scale) in the solvation layer were estimated using Eq. 36:

$$\gamma_2(solvation\ layer) = \frac{a_2}{Z_2^M} \tag{36}$$

where Z_2^M is the organic solvent mass fraction in the solvation layer; ρ_2^0 is the density of pure organic solvent:

$$Z_2^M = \frac{\rho_2}{\rho_2^0} w_2.$$

Eqs. 37-39 were used to calculate the partial densities of water (component 1) and the second component (organic solvent or protein) in the binary mixtures:

$$\rho^M = w_1\rho_1 + w_2\rho_2 \tag{37}$$

$$\rho_1 = \rho^M - w_2 \left(\frac{d\rho^M}{dw_2}\right)_{T,P} \tag{38}$$

$$\rho_2 = \rho^M - w_1 \left(\frac{d\rho^M}{dw_1}\right)_{T,P} \tag{39}$$

here ρ^M (g/cm³) is the density of binary water-organic or water-protein mixture; ρ_1 and ρ_2 are the partial densities of water and component 2, respectively. w_1 is the water mass fraction in binary mixtures; and w_2 is the mass fraction of component 2 in binary mixtures ($w_1 + w_2 = 1.0$).

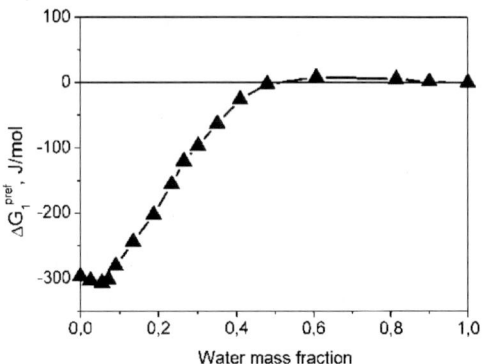

Figure 13. The Gibbs energies of the transfer of water (ΔG_1^{pref}) from the water-dimethyl sulfoxide (DMSO) mixtures to the solvation layer at 25°C.

The Gibbs energies of the transfer of water (ΔG_1^{pref}) from the water-dimethyl sulfoxide (DMSO) mixtures to the solvation layer are presented in Figure 13. The ΔG_1^{pref} values represent the actual thermodynamic situation of water in the solvation shell. ΔG_1^{pref} is a response of the solvation shell in terms of quantity G when an infinitesimal amount of water is added. It could thus signify the actual situation of water in the solvation layer in terms of quantity G.

The composition derivative (F_{i-i}^E) was proposed [63]:

$$F_{i-i}^E = (1 - x_i)\left(\frac{dF_i^E}{dx_i}\right) \qquad (40)$$

This quantity could be interpreted as the effect of incoming i on the thermodynamic situation of existing i in terms of F_i^E. Namely, F_{i-i}^E signifies the degree of the i-i interaction in terms of F^E. In the present study, we follow the same methodology.

The ΔG_{1-1}^{pref} values were estimated using Eq. 41:

$$\Delta G_{1-1}^{pref} = (1 - x_1)\left(\frac{d\Delta G_1^{pref}}{dx_1}\right) \qquad (41)$$

The resulting ΔG_{1-1}^{pref} data are given in Figure 14. As concluded from Figure 14, there are three mixing schemes in DMSO.

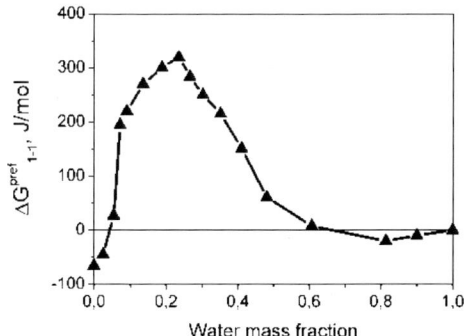

Figure 14. ΔG_{1-1}^{pref} as a function of water content in DMSO.

If composition derivative is positive, this implies the ΔG_{1-1}^{pref} contribution of water becomes more positive on introduction of more water to the solvation shell. Following the stability criterion, this means that the 1-1 interaction is unfavourable in terms of G.

Figure 14 shows that the water-water interaction in the solvation layer is attractive in terms of Gibbs energy at high ($w_1 = 0.6 - 1.0$) water content. On ther other hand, the water-water association in the solvation layer is unfavourable in terms of Gibbs energy at intermediate ($w_1 = 0.05 - 0.6$) water content. The water-water interaction in the solvation layer is attractive in terms of Gibbs energy at the lowest ($w_1 = 0 - 0.05$) water content. However, this effect is very weak.

These facts are in agreement with previously published results. These three regions were previously studied using thermodynamic and scattering methods [63]. In the water-rich region, a small number of DMSO molecules are bound together by S=O dipole attraction more tightly than in pure DMSO liquid. In solution, such a cluster enhances the hydrogen bond network of water in its immediate vicinity.

In the DMSO-rich region, DMSO molecules form clusters in exactly the same manner as in pure liquid DMSO, in which DMSO molecules are bound together not only by S=O dipole but also van der Waals interaction.

In the intermediate composition region, the solution consists of two kinds of clusters.

3.6. Effect of Preferentilal Interactions on the Enzyme Hydration and Activity

The sorption and enzyme activity experiments can be summarized as follows. Two concentration regimes were observed for water-DMSO mixtures:

(i) At high water content, lysozyme is in the native (preferentially hydrated) state. Lysozyme has a higher affinity for water than for DMSO. The ΔG_1^{pref}, ΔG_2^{pref}, and $(\partial g_3/\partial g_2)_{T,\mu_1,\mu_3}$ values are negative. On the other hand, the $(\partial g_1/\partial g_2)_{T,\mu_1,\mu_3}$ and R_1^E values are positive. Our conclusion agrees with previously published data for water-rich DMSO mixtures. From equilibrium dialysis experiments [2], it was shown that native protein is preferentially hydrated at room temperature. α-Chymotrypsin exhibits increased thermostability at high water content in DMSO [64].

(ii) At low and intermediate water content, the preferential binding of DMSO to lysozyme was detected. DMSO is a strong H-bond acceptor [50]. This means that DMSO is effective in solvating the dehydrated protein alone, in the absence of water. Considerable structural rearrangements and exothermic heat effects were observed in the solvents capable of forming strong hydrogen bonds (for example, DMSO) [55]. The ΔG_1^{pref}, ΔG_2^{pref}, and $(\partial g_3/\partial g_2)_{T,\mu_1,\mu_3}$ values are positive. On the other hand, the excess residual enzyme activity (R_1^E) values are negative. DMSO-induced irreversible inactivation was observed in this concentration region. DMSO may be considered as an informative tool for analyzing the effect of the hydrogen bond accepting ability on the protein stability in water-organic mixtures.

REFERENCES

[1] Timasheff, S. N. (2002). Protein-solvent preferential interactions, protein hydration, and the modulation of biochemical reactions by solvent components. *Proc. Natl. Acad. Sci. USA.*, *99*, 9721-9726.

[2] Arakawa, T., Kita, Y. & Timasheff, S. N. (2007). Protein precipitation and denaturation by dimethyl sulfoxide. *Biophys. Chem.*, *131*, 62-70.

[3] Gekko, K., Ohmae, E., Kameyama, K. & Takagi, T. (1998). Acetonitrile-protein interactions: Amino acid solubility and preferential solvation. *Biochim. Biophys. Acta*, *1387*, 195-205.

[4] Kovrigin, E. L. & Potekhin, S. A. (2000). On stabilizing action of protein denaturants: Acetonitrile effect on stability of lysozyme in aqueous solutions. *Biophys. Chem.*, *83*, 45-59.

[5] Shimizu, S. & Matubayasi, N. (2014). Preferential solvation: Dividing surface vs excess numbers. *J. Phys. Chem. B.*, *118*, 3922-3930.

[6] Auton, M., Bolen, D. W. & Rosgen, J. (2008). Structural thermodynamics of protein preferential solvation: Osmolyte solvation of proteins, amino acids, and peptides. *Proteins*, *73*, 802–813.

[7] Casassa, E. F. & Eisenberg, H. (1964). Thermodynamic analysis of multicomponent solutions. *Adv. Protein Chem.*, *19*, 287–395.

[8] Tanford, C. (1969). Extension of theory of linked functions to incorporate effects of protein hydration. *J. Mol. Biol.*, *39*, 539–544.

[9] Schellman, J. A. (1987). Selective binding and solvent denaturation. *Biopolymers*, *26*, 549−559.

[10] Arakawa, T., Bhat, R. & Timasheff, S. N. (1990). Why preferential hydration does not always stabilize the native structure of globular proteins. *Biochem.*, *29*, 1924-1931.

[11] Smith, P. E. (2006). Equilibrium dialysis data and the relationships between preferential interaction parameters in biological systems in terms of Kirkwood-Buff integrals. *J. Phys. Chem. B*, *110*, 2862− 2868.

[12] Parsegian, V. A., Rand, R. P. & Rau, D. C. (2000). Osmotic stress, crowding, preferential hydration, and binding: A comparison of perspectives. *Proc. Natl. Acad. Sci. U.S.A.*, *97*, 3987−3992.

[13] Kirby Hade, E. P. & Tanford, C. (1967). Isopiestic compositions as a measure of preferential interactions of macromolecules in two-component solvents. Application to proteins in concentrated aqueous cesium chloride and guanidine hydrochloride. *J. Amer. Chem. Soc.*, *89*, 5034-5040.

[14] Kamiyama, T., Liu, H. L. & Kimura, T. (2009). Preferential solvation of lysozyme by dimethyl sulfoxide in binary solutions of water and dimethyl sulfoxide. *J. Therm. Anal. Cal.*, *95*, 353-359.

[15] Reisler, E., Haik, Y. & Eisenberg, H. (1977). Bovine serum albumin in aqueous guanidine hydrochloride solutions. Preferential and absolute interactions and comparison with other systems. *Biochem.*, *16*, 197-203.

[16] Izumi, T., Yoshimura, Y. & Inoue, H. (1980). Solvation of lysozyme in water/dioxane mixtures studied in the frozen state by NMR spectroscopy. *Arch. Biophys. Biochem.*, *200*, 444-451.

[17] Gregory, R. B. (1995). Protein hydration and glass transition behavior. In: *Protein-Solvent Interactions* (Gregory, R.B., ed.), Marcel Dekker, New York, 191-264.

[18] Rupley, J. A. & Careri, G. (1991). Protein hydration and function. *Adv. Protein Chem.*, *41*, 37-172.

[19] Kuntz, I. D. & Kauzmann, W. (1974). Hydration of proteins and polypeptides. *Adv. Protein Chem.*, 28, 239-345.

[20] Oleinikova, A., Smolin, N., Brovchenko, I., Geiger, A. & Winter, R. (2005). Properties of spanning water networks at protein surfaces. *J. Phys. Chem. B.*, *109*, 1988-1998.

[21] Durchschlag, H. & Zipper, P. (2001). Comparative investigations of biopolymer hydration by physicochemical and modeling techniques. *Biophys. Chem.*, *93*, 141-157.

[22] Sirotkin, V. A. & Khadiullina, A. V. (2013). Gibbs energies, enthalpies, and entropies of water and lysozyme at the inner edge of excess hydration. *J. Chem. Phys.*, *139*, 075102/1-075102/9.

[23] Sirotkin, V. A. & Khadiullina, A. V. (2014). A study of the hydration of ribonuclease A using densitometry: Effect of the protein hydrophobicity and polarity. *Chem. Phys. Lett*, *603*, 13-17.

[24] Privalov, P. L. & Crane-Robinson, C. (2017). Role of water in the formation of macromolecular structures. *Eur. Biophys. J.*, *46*, 203-224.

[25] Bull, H. B. (1944). Adsorption of water vapor by proteins. *J. Amer. Chem. Soc.*, 66, 1499-1507.

[26] Luscher-Mattli, M. & Ruegg, M. (1982). Thermodynamic functions of biopolymer hydration. I. Their determination by vapor pressure studies, discussed in an analysis of the primary hydration process. *Biopolymers*, 21, 403-418.

[27] Luscher-Mattli, M. & Ruegg, M. (1982). Thermodynamic functions of biopolymer hydration. II. Enthalpy-entropy compensation in hydrophilic hydration process. *Biopolymers*, 21, 419-429.

[28] Bone, S. (1987). Time-domain reflectometry studies of water binding and structural flexibility in chymotrypsin. *Biochem. Biophys. Acta.*, *916*, 128-134.

[29] Sirotkin, V. A. & Khadiullina, A. V. (2011). Hydration of proteins: excess partial enthalpies of water and proteins. *J. Phys. Chem. B*, *115*, 15110-15118.

[30] Sirotkin, V. A., Komissarov, I. A. & Khadiullina, A. V. (2012). Hydration of proteins: excess partial volumes of water and proteins, *J. Phys. Chem. B*, *116*, 4098-4105.

[31] Sirotkin, V. A. (2005). Effect of dioxane on the structure and hydration-dehydration of α-chymotrypsin as measured by FTIR spectroscopy. *Biochim. Biophys. Acta.*, *1750*, 17-29.

[32] Klibanov, A. M. (2001). Improving enzymes by using them in organic solvents, *Nature*, *409*, 241-246.

[33] Carrea, G. & Riva, S. (2000). Properties and synthetic applications of enzymes in organic solvents, *Angew. Chem. Int. Ed.*, *39*, 2226-2254.

[34] Halling, P. J. (2004). What can we learn by studying enzymes in nonaqueous media? *Phil. Trans. R. Soc. Lond. B Biol. Sci.*, *359*, 1287–1297.

[35] Micaelo, N. M. & Soares C. M. (2007). Modeling hydration mechanisms of enzymes in nonpolar and polar organic solvents, *FEBS J.*, *274*, 2424–2436.

[36] Clark, D. S. (2004). Characteristics of nearly dry enzymes in organic solvents: implications for biocatalysis in the absence of water, *Phil. Trans. R. Soc. Lond. B Biol. Sci.*, *359*, 1299–1307.

[37] Serdakowski, A. L. & Dordick, J. S. (2007). Enzyme activation for organic solvents made easy, *Trends Biotechnol.*, *26*, 54-48.

[38] Rariy, R. & Klibanov, A. M. (1997) Correct protein folding in glycerol, *Proc. Natl. Acad. Sci. U.S.A.*, *94*, 13520-13523.

[39] Sirotkin, V. A., Hüttl, R. & Wolf, G., (2008). Enzyme-catalysed hydrolysis of L-amino acid esters in a low water organic solvent

studied by isothermal calorimetry, *J. Therm. Anal. Calorim*, *93*, 515-520.

[40] Sirotkin, V. A. & Faizullin, D. A. (2004). Interaction enthalpies of solid human serum albumin with water-dioxane mixtures: comparison with water and organic solvent vapor sorption, *Thermochim. Acta*, *415*, 127-133.

[41] Partridge, J., Hutcheon, G. A., Moore, B. D. & Halling P. J. (1996). Exploiting hydration hysteresis for high activity of cross-linked subtilisin crystals in acetonitrile, *J. Amer. Chem. Soc.*, *118*, 12873-12877.

[42] Borisover, M. D., Sirotkin, V. A. & Solomonov, B. N. (1995). Isotherm of water sorption by human serum albumin in dioxane: Comparison with calorimetric data. *J. Phys. Org. Chem.*, *8*, 84-88.

[43] Sirotkin, V. A., Borisover, M. D. & Solomonov, B. N. (1995). Heat effects and water sorption by human serum albumin on its suspension in water-dimethyl sulfoxide mixtures. *Thermochim. Acta*, *256*, 175-183.

[44] McMinn, J. H., Sowa, M. J., Charnick, S. B. & Paulaitis, M. E. (1993). The hydration of proteins in nearly anhydrous organic solvent suspensions, *Biopolymers.*, 33, 1213-1224.

[45] Kijima, T., Yamamoto, S. & Kise, H. (1996). Study of tryptophan fluorescence and catalytic activity of α-chymotrypsin in aqueous organic media, *Enz. Microb. Technol.*, *18*, 2-6.

[46] Khmelnitsky, Yu., Mozhaev, V. V., Belova, A. B., Sergeeva, M. V. & Martinek, K. (1991). Denaturation capacity: a new quantitative criterion for selection of organic solvents as reaction media in biocatalysis, *Eur. J. Biochem.*, *198*, 31-41.

[47] Simon, L. M., Kotorman, M., Garab, G. & Laczko, I. (2001). Structure and activity of δ-chymotrypsin and trypsin in aqueous organic media, *Biochem. Biophys. Res. Comm.*, *280*, 1367-1371.

[48] Fersht, A. (1999). *Structure and Mechanism in Protein Science: A Guide to Enzyme Catalysis and Protein Folding*; Freeman & Co: New York.

[49] Lehninger, A. L., Nelson, D. L. & Cox, M. M. (1993). *Principles of Biochemistry*; Worth: New York.

[50] Borisover, M. D., Stolov, A. A., Cherkasov, A. R., Izosimova, S. V. & Solomonov, B. N. (1994). Calorimetric and infrared spectroscopic study of intermolecular interactions of water in organic solvents, *Russ. J. Phys. Chem.*, *68*, 48-53.

[51] Perrin, D. D., Armarego, W. L. F. & Perrin, D. R. (1980). *Purification of Laboratory Chemicals*, Oxford: Pergamon Press.

[52] Borisover, M. D., Sirotkin, V. A. & Solomonov, B. N. (1995). Thermodynamics of water binding by human serum albumin suspended in acetonitrile. *Thermochim. Acta*, *254*, 47-53.

[53] Sirotkin, V. A. & Kuchierskaya, A. A. (2017). Preferential solvation/hydration of α-chymotrypsin in water-acetonitrile mixtures. *J. Phys. Chem. B.*, *121*, 4422-4430.

[54] Sirotkin, V. A. & Kuchierskaya, A. A. (2017). Lysozyme in water-acetonitrile mixtures: Preferential solvation at the inner edge of excess hydration. *J. Chem. Phys.*, *146*, 215101-8.

[55] Sirotkin, V. A. & Kuchierskaya, A. A. (2017). α-Chymotrypsin in water-ethanol mixtures: Effect of preferential interactions. *Chem. Phys. Lett.*, *689*, 156-161.

[56] Sirotkin, V. A. & Kuchierskaya, A. A. (2017). α-Chymotrypsin in water-acetone and water-dimethyl sulfoxide mixtures: Effect of preferential solvation and hydration. *Proteins: Functions, Structure and Bioinformatics*, *85*, 1808-1819.

[57] Sirotkin, V. A., Zinatullin, A. N., Solomonov, B. N., Faizullin, D. A. & Fedotov, V. D. (2002). Interaction enthalpies of solid bovine pancreatic chymotrypsin with organic solvents: comparison with FTIR-spectroscopic data. *Thermochim. Acta*, *382*, 151-160.

[58] Atkins, P. W. (2006). *Physical Chemistry*. 8th ed. Oxford: Oxford University Press.

[59] Prausnitz, J. M. (1969). *Molecular Thermodynamics of Fluid-Phase Equilibria*. N.J.: Prentice-Hall, Inc., Engelwood Cliffs.

[60] Belousov, V. P. & Panov, M. Y. (1994). *Thermodynamic properties of aqueous solutions of organic substances*. Boca Raton, Fla.: CRC Press.

[61] Bell, G., Janssen, A. E. M. & Halling, P. (1996). Water activity fails to predict critical hydration level for enzyme activity in polar organic solvents: Interconversion of water concentrations and activities. *Enzym. Microb. Technol.*, *20*, 471-476.

[62] Pendin, A. A. (1989). Preferential solvation and thermodynamical properties of nonelectrolites solutions. *Russ, J. Phys. Chem.* 63, 1793-1798.

[63] Lai, J. T. W., Lau, F. W., Robb, D., Westh, P., Nielsen, G., Trandum, C., Hvidt, A. Koga, Y. (1995). Excess partial molar enthalpies, entropies, Gibbs energies, and volumes in aqueous dimethylsulfoxide. *J. Sol. Chem. 24,* 89-102.

[64] Tretyakova, T., Shushanyan. M., Partskhaladze, T., Makharadze. M., van Eldik R. & Khoshtariya, D. E. (2013). Simplicity within the complexity: Bilateral impact of DMSO on the functional and unfolding patterns of α-chymotrypsin. *Biophys. Chem.*, *175–176*, 17–27.

Chapter 5

PREFERENTIAL SOLVATION AND HYDRATION OF LYSOZYME IN ETHYLENE GLYCOL AND ETHANOL: EFFECT OF HYDROXYL GROUP

ABSTRACT

A thermodynamic description of the preferential solvation and preferential hydration of hen egg-white lysozyme in water-alcohol (ethanol and ethylene glycol) mixtures was performed. Residual enzyme activity and absolute values of the water/alcohol sorption were investigated at 25°C. One of the most important advantages of our approach is the facilitation of individual evaluation of the Gibbs energies of water, alcohol, and protein at low water content. There are three concentration regimes for ethanol. Protein is preferentially hydrated at high water content. The residual enzyme activity values are close to 100%. The dried enzyme has a higher affinity for alcohol than for water at intermediate water content. Residual enzyme activity is minimal in this concentration range. The ethanol molecules are preferentially excluded from the protein surface at the lowest water content. This results in preferential hydration of lysozyme. The residual catalytic activity is ~60% in water-poor ethanol. Opposite to ethanol, the residual activity is close to zero in water-poor ethylene glycol. Our data clearly demonstrate that the replacement of the hydroxyl group by methyl group constitutes a critical factor in determining the stability of protein-water-alcohol systems.

Keywords: preferential solvation, preferential hydration, lysozyme, water, alcohol, ethanol, ethylene glycol

1. INTRODUCTION

There are numerous investigations which demonstrate the bilateral action of monohydric alcohols on the protein properties [1-6]. For example, the temperature of protein denaturation in monohydric alcohols decreases gradually with augmenting organic solvent concentration [1-4]. This effect becomes more pronounced with increasing length of the alkyl chain. On the other hand, the denaturation enthalpy passes through a maximum with augmenting alcohol content [2-6]. At low alcohol content and temperatures around 0-25°C, monohydric alcohols can slightly stabilize proteins [4-6].

Understanding the bilateral impact of monohydric alcohols on the protein stability requires effective techniques that reveal biophysical information regarding protein–alcohol and protein–water interactions. Preferential solvation/hydration may be an effective and informative approach for elucidating the dual effect of water-alcohol mixtures on the protein stability. Preferential solvation is a thermodynamic quantity that describes the protein occupancy by the alcohol/water molecules [7-22]. Alcohol and water exist preferentially in the solvation layer of the protein. When a protein is placed into a water-alcohol mixture, its properties are altered as a function of the solvent composition. The preferential solvation/hydration process accounts for the augmentation or depletion of the alcohol/water molecules at the protein surface. Preferential binding is the excess of alcohol at the protein surface relative to the alcohol content in the bulk solvent. The preferential binding depends markedly on the chemical nature of the protein surface. For example, protein unfolding may be induced by the preferential binding to specific regions on the protein (to peptide groups in the case of urea and guanidinium hydrochloride or to hydrophobic regions in the case of alcohols) [7, 8, 18-22].

Polyols (polyhydric alcohols) prevent the loss of enzyme activity and increase the temperature/enthalpy of protein denaturation. It was concluded [7, 8, 11, 12, 14, 15, 16] that the protein stabilization by polyols is due to a preferential hydration (preferential exclusion) effect.

Water binding (hydration or biological water) plays a key role in determining the structure, stability, dynamics, and functions of proteins [23-37]. Therefore, the aim of this study is to compare the preferential solvation and preferential hydration of protein macromolecules at low, intermediate, and high water content in monohydric and polyhydric alcohols at 25°C. Our approach is based on the analysis of absolute values of the alcohol/water sorption and residual enzyme activity. One of the most important advantages of our approach is the facilitation of individual evaluation of the Gibbs energies of water, alcohol, and protein at low water content.

Hen egg-white lysozyme was used as a model protein. This protein is one of the most applied and studied in biophysical and biotechnological investigations [38, 39]. Lysozyme is a small monomeric protein of 129 amino acid residues. The physiological role of lysozyme is to hydrolyze polysaccharide chains [38, 39].

The choice of ethanol (EtOH) and ethylene glycol (EG) was determined for the following reasons:

A) Ethanol and ethylene glycol are water-miscible organic solvents. Therefore, it is possible to study the effect of these substances on the functions and hydration/solvation of lysozyme in the entire range of water content.

B) Ethanol and ethylene glycol have similar chemical structures. The replacement effect of OH from EG by H (resulting in ethanol) may be studied. For example, the water activity coefficients in EG are lower than one (negative deviations from the ideality). On the other hand, the water activity coefficients in ethanol are higher than one (positive deviations from the ideality).

C) Ethylene glycol has a stabilizing effect on the native structure of proteins. There are numerous observations which demonstrate the stabilizing effect of EG. For example, the denaturation temperature and enthalpy of proteins (lysozyme, ribonuclease) are higher in water-EG mixtures that in pure aqueous solution [1-5]. The stabilization of lysozyme and ribonuclease in the presence of EG has been also shown [1-5].

This means that these organic liquids may be considered as an informative tool for analyzing the effect of replacement of the hydroxyl group on preferential interactions in the protein-water-alcohol systems.

2. EXPERIMENTAL

2.1. Materials

Hen egg-white lysozyme (EC 3.2.1.17; crystallized three times, dialyzed, and lyophilized and dried *Micrococcus lysodeikticus* cells were purchased from Sigma Chemical Co. (St. Louis, MO, U.S.A.). The molecular weight of the protein was taken as 14300 Da. Ethanol and ethylene glycol (analytical grade, with a purity of > 99%) were purified and dried according to the recommended guidelines [40]. Water used was doubly distilled. All water-organic mixtures were prepared gravimetrically using a Precisa balance (Swiss) with a precision of 0.00001 g.

2.2. Initial Protein State

The lysozyme powder was placed on the thermostated cell and dried using a microthermoanalyzer "Setaram" MGDTD-17S (±0.00001 g) at 25°C and 0.1 Pa, until a constant sample weight was reached. The water content of the dehydrated protein was estimated as 0.002 ± 0.001 g water/g protein, using the Karl Fischer titration method as described previously [41-44]. This value for lysozyme implies that at the *zero hydration level* there are about two water molecules strongly bound to each protein molecule.

2.3. Sorption Measurements

The protein samples were prepared according to the recommended guidelines [41, 42]. The initially dried protein samples were presented to water-alcohol vapor mixtures. The water-organic vapor mixture was consecutively flowed through a thermostated saturator filled with the water-alcohol mixture, and a cell containing the protein sample. Protein samples (5-10 mg) each were flushed by water – alcohol vapor mixtures, until no further mass changes were detected as described previously. The sorption equilibrium was reached after 6 h at 25°C. The schematic representation of the experimental setup is given in Ref. 41. An external ethylene glycol thermostat (RC 6, Lauda, Germany) was utilized to determine the temperature with a precision of 0.1°C. The water activity (a_1) in the vapor phase was adjusted by altering the water content in the liquid water-alcohol mixture.

Measurements of the water sorption (A_1) were conducted by Karl Fischer titration with a Metrohm 831 KF coulometer. Alcohol content of lysozyme (A_3) was calculated as a difference between the total sorption uptake (A_1+A_3) and water content (A_1). The total sorption uptake (A_1+A_3) was measured by microthermoanalyzer "Setaram" MGDTD-17S.

2.4. Residual Enzyme Activity

Residual enzyme activity was determined by measuring the enzyme activity after storage in water-alcohol mixtures according to the recommended guidelines [41, 42]. The lysozyme activity was determined as follows. The lysozyme sample was immersed in an aqueous-alcohol mixture of required composition and was incubated at 25°C for 3 h. This time period exceeded the time corresponding to the completion of the calorimetric heat effect accompanying the interaction of the dried proteins with pure organic solvents and water-organic mixtures [45]. The concentration of lysozyme in the water-alcohol mixtures was 1 mg/ml. Adding 100-μl aliquots of the lysozyme solution in the water-rich ethanol/EG (or the lysozyme suspension in the water-poor ethanol/EG) to the aqueous solution of the substrate (*Micrococcus lysodeikticus* cells (2.9 ml, 0.3 mg/ml) in 0.1 M potassium phosphate pH 7.0), we initiated the enzymatic reaction. Change in the absorbance at 450 nm was detected using a Perkin-Elmer Lambda 35 double-beam scanning spectrophotometer. The reaction was followed for 300-1500 s. Each kinetic curve was reproduced not less than three times.

3. RESULTS AND DISCUSSION

3.1. Sorption Isotherms. Excess Sorption Functions

Figures 1 and 2 present the water (A_1) and alcohol (A_3) vapor sorption isotherms for lysozyme at 25°C. The lysozyme solvation layer is composed of two parts: (i) non-ideal (due to preferential interactions) and (ii) ideal. The non-ideal effect of the solvation layer on enzyme properties (alcohol/water sorption; residual enzyme activity) can be defined in terms of excess functions, F^E [46-48], which refers to the difference between the observed function of mixing, F^M, and the function for an ideal binary mixture, F^M_{id}.

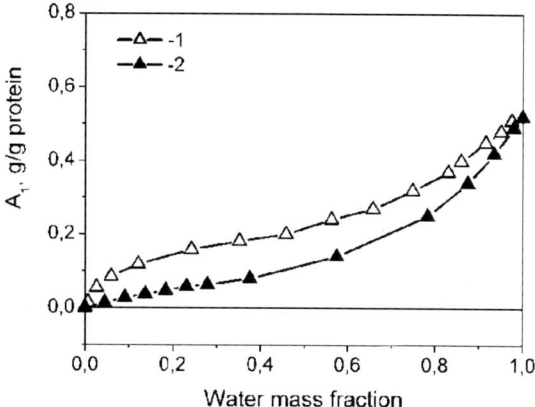

Figure 1. Water (A_1) sorption isotherms for lysozyme at 25°C: 1 – Ethanol; 2 – Ethylene glycol. The standard errors of estimation of the water sorption were 0.001-0.002 g/g. Each experiment was performed 3-4 times.

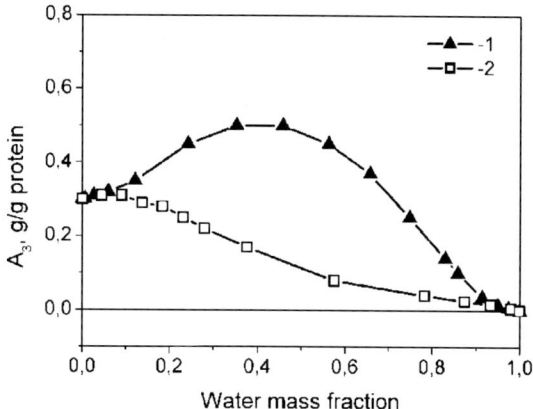

Figure 2. Alcohol (A_3) sorption isotherms for lysozyme at 25°C: 1 – Ethanol; 2 – Ethylene glycol. The standard errors of estimation of the organic solvent sorption were 0.001-0.002 g/g. Each experiment was performed 3-4 times.

The deviations of the excess functions from zero describe the degree to which the protein solvation layer differs from the pure binary water-alcohol mixture as a consequence of preferential interactions between water (component 1), protein (component 2), and alcohol (component 3).

Eq. 1 is used to calculate the F^E values:

$$F^E = F^M - F_{id}^M \tag{1}$$

The $F_{id,i}^M$ values were calculated using Eq. 2:

$$F_{id,i}^M = F_i^M(w_i = 0) + w_i[F_i^M(w_i = 1.0) - F_i^M(w_i = 0)] \tag{2}$$

where $F_i^M(w_i = 1.0)$ is the observed mixing function of lysozyme at $w_i = 1.0$; $F_i^M(w_i = 0)$ is the observed mixing function of lysozyme at $w_i = 0$; w_1 is the water mass fraction in the binary water-organic mixtures; and w_3 is the organic solvent mass fraction in the binary water-organic mixtures ($w_1 + w_3 = 1.0$).

The situation in which no preferential interactions exist between water, protein, and alcohol is described by the $F_{id,i}^M$ values. In this circumstance, the water mass fraction in the ideal part of the solvation layer is the same as in the pure water-alcohol mixture.

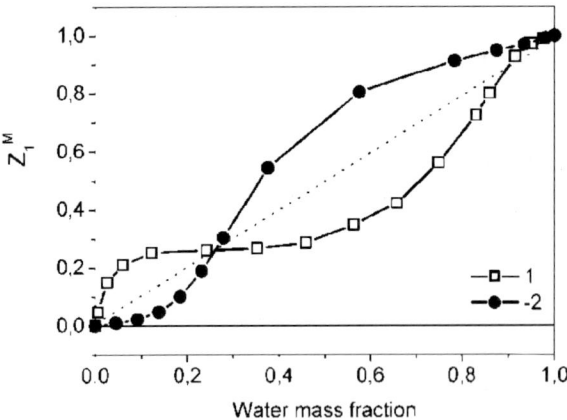

Figure 3. Water mass fraction in the solvation layer of lysozyme (Z_1^M): 1 – Ethanol; 2 – Ethylene glycol.

The Z_1^M (water mass fraction in the solvation layer) and Z_3^M (alcohol mass fraction in the solvation layer) values as a function of water mass fraction in alcohol are presented in Figure 3. Eqs. 3 and 4 are utilized to calculate the Z_1^M and Z_3^M values, respectively:

$$Z_1^M = \frac{A_1}{A_1 + A_3} \tag{3}$$

$$Z_3^M = \frac{A_3}{A_1 + A_3} \tag{4}$$

The simultaneous action of alcohol and water was determined by the Z_1^E and Z_3^E values (Figures 4 and 5). These excess sorption functions can be estimated using Eqs. 5 and 6:

$$Z_1^E = Z_1^M - Z_{id,1}^M \tag{5}$$

$$Z_3^E = Z_3^M - Z_{id,3}^M \tag{6}$$

where Z_1^M is the mass fraction of water in the solvation layer for the real water-alcohol mixture; and $Z_{id,1}^M$ is the mass fraction of water in the solvation layer for the ideal water-alcohol mixture.

The $Z_{id,1}^M$ values were calculated using Eq. 7:

$$Z_{id,1}^M = Z_1^M(w_1 = 0) + w_1[Z_1^M(w_1 = 1.0) - Z_1^M(w_1 = 0)] \tag{7}$$

where $Z_1^M(w_1 = 1.0)$ is the water mass fraction in the solvation layer of lysozyme at $w_1=1.0$; $Z_1^M(w_1 = 0)$ is the water mass fraction in the solvation shell at $w_1=0$; and w_1 is the mass fraction of water in alcohol.

In addition, Z_3^M is the alcohol mass fraction in the solvation layer for the real water-alcohol mixture; and $Z_{id,3}^M$ is the alcohol mass fraction for the ideal water-alcohol mixture. Eq. 8 is used to calculate the $Z_{id,3}^M$ values:

$$Z_{id,3}^M = Z_3^M(w_3 = 0) + w_3[Z_3^M(w_3 = 1.0) - Z_3^M(w_3 = 0)] \qquad (8)$$

where $Z_3^M(w_3 = 0)$ is the alcohol mass fraction in the solvation shell of lysozyme at $w_3=1.0$; $Z_3^M(w_3 = 0)$ is the alcohol mass fraction in the solvation shell of lysozyme at $w_3=0$; and w_3 is the alcohol mass fraction in the binary water-organic mixture.

3.1.1. Ethanol

As shown in Figure 4, the Z_1^E values are positive at low ($w_1 \sim 0\text{-}0.2$) and high ($w_1 \sim 0.9\text{-}1.0$) water content. A considerable decrease in the water sorption was detected at intermediate water content. The Z_1^E values are negative in this concentration interval. The most pronounced decrease was found in the water mass fraction range from 0.4 to 0.8.

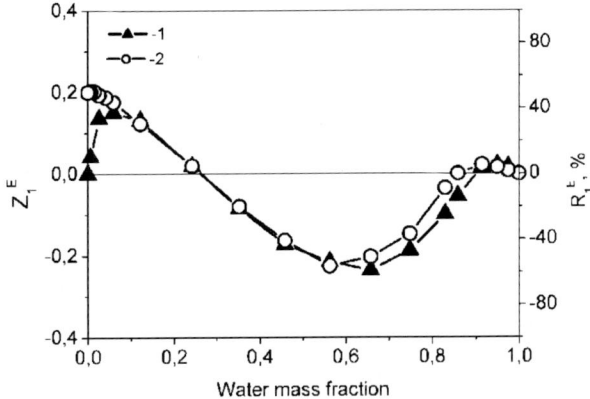

Figure 4. Excess water mass fraction in the solvation layer of lysozyme in water-ethanol mixtures (Z_1^E). (2) Excess residual activity of lysozyme in water-ethanol mixtures (R_1^E).

3.1.2. Ethylene Glycol

Figure 5 shows that the Z_1^E values are positive at high ($w_1 = 0.7\text{-}1.0$) water content. The Z_1^E values are negative at low water content ($w_1 = 0\text{-}0.4$).

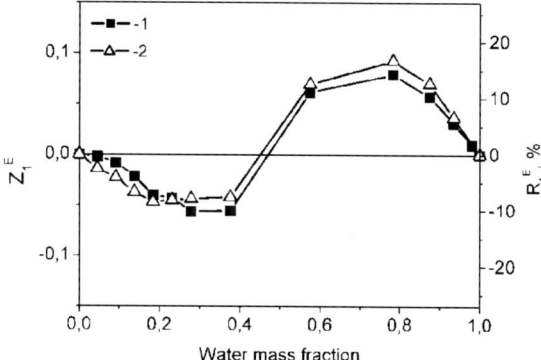

Figure 5. Excess water mass fraction in the solvation layer of lysozyme in water-ethylene glycol mixtures (Z_1^E). (2) Excess residual activity of lysozyme in water-ethylene glycol mixtures (R_1^E).

3.2. Residual Enzyme Activity

Figure 6 demonstrates typical kinetic curves for the enzymatic reaction catalyzed by lysozyme preliminarily incubated in water-alcohol mixtures. The catalytic activity was characterized by the ratio of the extent of hydrolysis attained within 300 s with lysozyme incubated in a water-alcohol mixture to the same quantity measured using lysozyme incubated in pure water (Figure 6, curve 1).

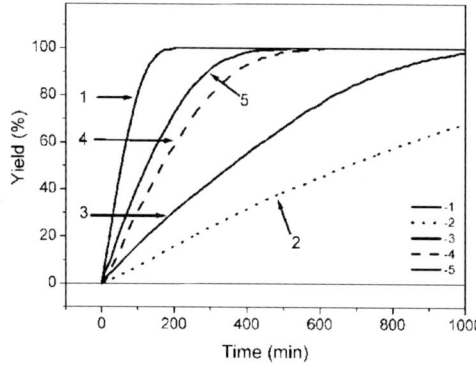

Figure 6. Typical kinetic curves for the enzymatic reaction catalyzed by lysozyme previously incubated in water-alcohol mixtures. Ethanol. Water mass fraction in ythanol. (1) 1.0; (2) 0.66; (3) 0.26; (4) 0.05; (5) 0,82.

The residual activity values are shown in Figure 7. As shown in Figure 7, alcohols affect the residual enzyme activity of lysozyme in a complicated manner.

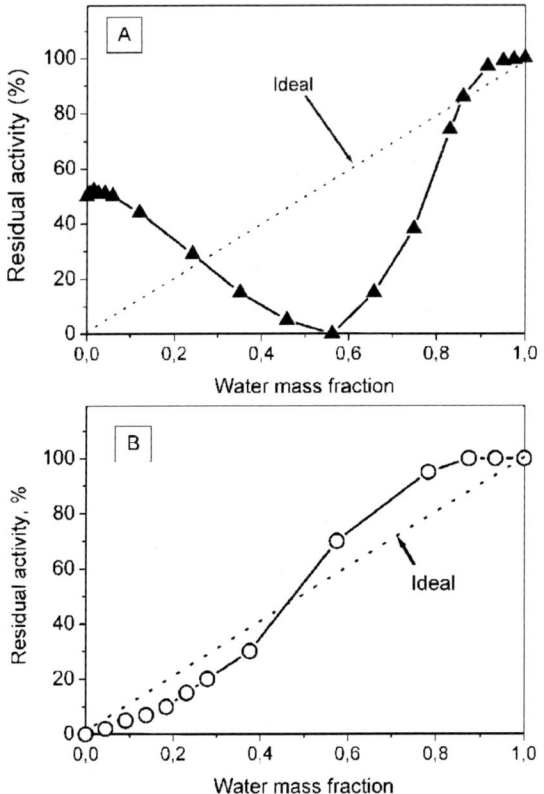

Figure 7. Residual activity of lysozyme in water-alcohol mixtures: (A) –Ethanol; (B) – Ethylene glycol. All values are the averages of three measurements. Experimental errors were 1-2%.

3.2.1. Ethanol

The residual activity values are close to 100% (Figure 7A) at high water content (w_1~0.9-1.0). There is a sharp transition from the water-rich range to the intermediate one at $w_1 < 0.9$. The residual enzyme activity of lysozyme changes from 100 to 0% in the transition region (w_1~0.6-0.8). A minimum on the residual activity curve was found at

$w_1 \sim 0.6$ in ethanol. The residual activity augments at $w_1 < 0.4$. The residual enzyme activity is ~50% in water-poor ethanol.

3.2.2. Ethylene Glycol

The residual activity values are close to 100% (Figure 7B), at high water content ($w_1 \sim 0.8$-1.0). The residual enzyme activity alters from 100 to 0% in the transition range ($w_1 \sim 0.2$-0.8). At low water content ($w_1 < 0.1$), the residual activity values are close to zero.

3.3. Excess Residual Enzyme Activity

The effect of the excess hydration (Z_1^E) of the residual enzyme activity was described by the R_1^E values (excess residual enzyme activity) (Figures. 4 and 5). The R_1^E values were estimated using Eq. 9:

$$R_1^E = R^M - R_{id,1}^M \qquad (9)$$

where R^M is the observed residual enzyme activity; and $R_{id,1}^M$ is the function for an ideal binary mixture.

The R_{id}^M values can be calculated using Eq. 10:

$$R_{id,1}^M = R^M(w_1 = 0) + w_1[R^M(w_1 = 1.0) - R^M(w_1 = 0)] \qquad (10)$$

where $R^M(w_1 = 1.0)$ is the residual activity of lysozyme at $w_1 = 1.0$; $R^M(w_1 = 0)$ is the residual activity of lysozyme at $w_1 = 0$; w_1 is the water mass fraction in the water-alcohol mixtures; and w_3 is the alcohol mass fraction in the water-organic mixtures ($w_1 + w_3 = 1.0$).

Figures 4 and 5 demonstrate the dependencies of the R_1^E values on the water concentration in alcohols. As concluded from Figures 4 and 5, the R_1^E values differ significantly from zero. This means that the effect of the water-alcohol solvation shell on the residual enzyme activity is non-ideal in the entire range of water content. It is worth noting that the R_1^E values are consistent with the Z_1^E values (Figures 4 and 5).

Three concentration regions were found for ethanol (Figure 4):

(i) The R_1^E and Z_1^E values are positive at high water content.
(ii) At intermediate water content, the R_1^E and Z_1^E values are negative. Ethanol augments the degree of irreversible inactivation of lysozyme in this region.
(iii) At low water level, the R_1^E and Z_1^E values are positive in water-poor ethanol.

Two different regimes are operative for ethylene glycol (Figure 5):

(i) The $\boldsymbol{R_1^E}$ and $\boldsymbol{Z_1^E}$ values are positive at intermediate and high water content.
(ii) Opposite to ethanol, the $\boldsymbol{R_1^E}$ and $\boldsymbol{Z_1^E}$ values are very small for the water-poor EG. Ethylene glycol induces irreversible enzyme inactivation at low water content.

3.4. Preferential Solvation Parameters

The degree to which the solvation layer differs from the pure binary water-alcohol mixture due to preferential interactions between water (component 1), protein (component 2), and alcohol (component 3) was characterized by the preferential interaction parameters (Eqs. 11 and 12). Eq. 11 [7-9] was utilized to determine the preferential binding of lysozyme in water-alcohol mixtures:

$$(\partial g_3 / \partial g_2)_{T, \mu_1, \mu_3} = A_3 - \frac{w_3}{w_1} A_1 \qquad (11)$$

where A_1 is the protein hydration, expressed as gram water per gram protein; A_3 is the binding of alcohol, expressed as gram organic solvent per gram protein; w_1 is the water mass fraction in water-alcohol mixtures; and w_3 is the mass fraction of organic solvent in water-alcohol mixtures ($w_1 + w_3 = 1.0$).

Eq. 12 was used to estimate the preferential hydration:

$$(\partial g_1/\partial g_2)_{T,\mu_1,\mu_3} = -\left(\frac{w_1}{w_3}\right)(\partial g_3/\partial g_2)_{T,\mu_1,\mu_3} \tag{12}$$

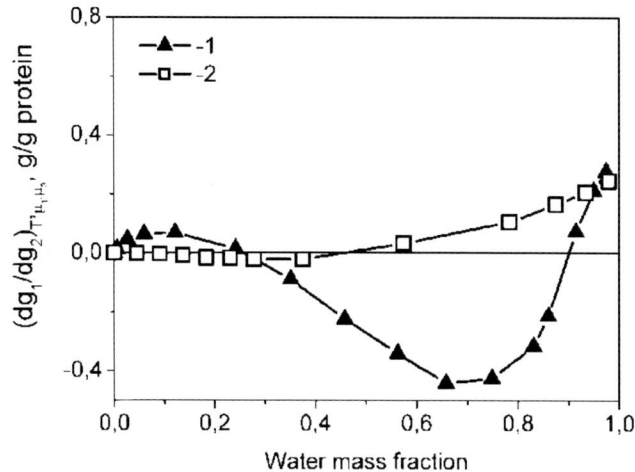

Figure 8. The preferential hydration parameters of lysozyme as a function of water mass fraction in alcohols (($\partial g_1/\partial g_2)_{T,\mu_1,\mu_3}$): 1 – Ethanol; 2 – Ethylene glycol.

Figures 8 and 10 show the preferential interaction parameters calculated using Eqs. 11 and 12, respectively.

Eqs. 13 and 14 were utilized to determine the Gibbs energies of the transfer of water (ΔG_1^{pref}) and alcohol (ΔG_3^{pref}) from water-alcohol mixtures to the solvation layer of lysozyme:

$$\Delta G_1^{pref} = \mu_1^E(solvation\ layer) - \mu_1^E(binary\ mixture) \tag{13}$$

$$\Delta G_3^{pref} = \mu_3^E(solvation\ layer) - \mu_3^E(binary\ mixture) \qquad (14)$$

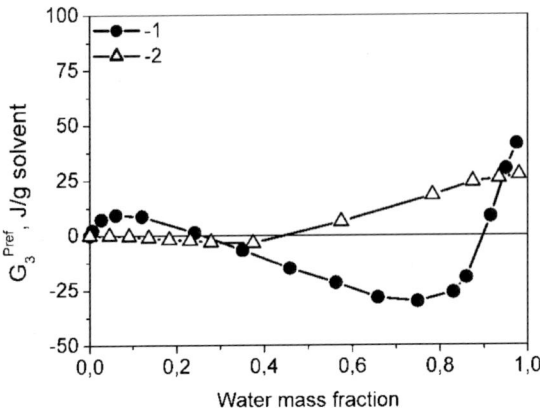

Figure 9. Gibbs energy of the transfer of alcohols (ΔG_3^{pref}) from binary water-organic mixtures to the solvation shell of lysozyme: 1 – Ethanol; 2 – Ethylene glycol.

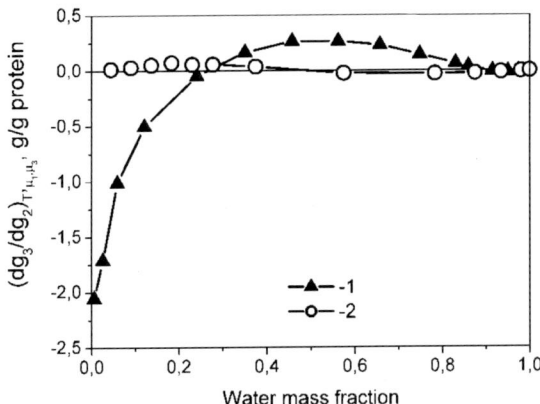

Figure 10. The preferential solvation parameters of lysozyme as a function of water mass fraction in alcohols (($\partial g_3/\partial g_2)_{T,\mu_1,\mu_3}$): 1 – Ethanol; 2 – Ethylene glycol.

The $\mu_1^E(binary\ mixture)$ and $\mu_3^E(binary\ mixture)$ values can be calculated using Eqs. 15 and 16:

$$\mu_1^E(binary\ mixture) = RT ln\gamma_1(binary\ mixture) \qquad (15)$$

$$\mu_3^E(binary\ mixture) = RTln\gamma_3(binary\ mixture) \qquad (16)$$

Water activity coefficients ($\gamma_1(binary\ mixture)$, the mass fraction scale; the reference state is pure water) in aqueous-organic mixtures were estimated using Eq. 17:

$$\gamma_1(binary\ mixture) = \frac{a_1}{w_1} \qquad (17)$$

Alcohol activity coefficients ($\gamma_3(binary\ mixture)$, the mass fraction scale; the reference state is pure alcohol) in aqueous-organic mixtures were calculated using Eq. 18:

$$\gamma_3(binary\ mixture) = \frac{a_3}{w_3} \qquad (18)$$

Water activity (a_1) and alcohol activity (a_3) were taken from the published data [49] based on the vapor-liquid equilibrium.

The $\mu_1^E(solvation\ layer)$ and $\mu_3^E(solvation\ layer)$ values were calculated using Eqs. 19 and 20:

$$\mu_1^E(solvation\ layer) = RTln\gamma_1(solvation\ layer) \qquad (19)$$

$$\mu_3^E(solvation\ layer) = RTln\gamma_3(solvation\ layer) \qquad (20)$$

Water activity coefficients ($\gamma_1(solvation\ layer)$, the mass fraction scale) in the solvation layer were calculated using Eq. 21:

$$\gamma_1(solvation\ layer) = \frac{a_1}{Z_1^M} \qquad (21)$$

where Z_1^M is the mass fraction of water in the solvation layer;

$$Z_1^M = \frac{A_1}{A_1+A_3}.$$

Alcohol activity coefficients (γ_3(*solvation layer*), the mass fraction scale) in the solvation layer were estimated using Eq. 22:

$$\gamma_3(solvation\ layer) = \frac{a_3}{Z_3^M} \qquad (22)$$

where Z_3^M is the alcohol mass fraction in the solvation layer;

$$Z_3^M = \frac{A_3}{A_1+A_3}.$$

Figures 9 and 11 show the ΔG_3^{pref} and ΔG_1^{pref} values. Figures 7-10 demonstrate that the ΔG_3^{pref} and ΔG_1^{pref} values correlate well with the preferential interaction parameters.

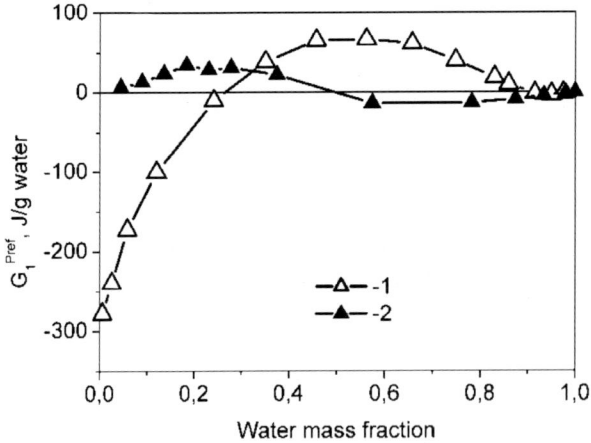

Figure 11. Gibbs energy of the transfer of water (ΔG_1^{pref}) from binary water-organic mixtures to the solvation shell of lysozyme: 1 – Ethanol; 2 - Ethylene glycol.

ΔG_2^{pref} values (Figure 12) described lysozyme stabilization and destabilization due to the preferential interactions. The Gibbs-Duhem equation for ternary systems (Eq. 23) was utilized to calculate the ΔG_2^{pref} values:

$$\Delta G_2^{pref} = \frac{m_1 \Delta G_1^{pref} + m_3 \Delta G_3^{pref}}{m_2} \qquad (23)$$

$$\Delta G_2^{pref} = \mu_2^E(\text{protein in water} - \text{alcohol mixtures}) - \mu_2^E(\text{pure protein}) \qquad (24)$$

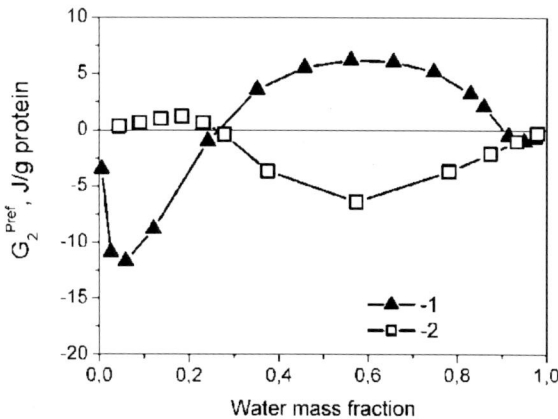

Figure 12. Gibbs energy (ΔG_2^{pref}) of the transfer of lysozyme from pure state to water-alcohol mixtures: 1 – Ethanol; 2 – Ethylene glycol.

where μ_1^E, μ_2^E, and μ_3^E are the excess chemical potentials of water, protein, and alcohol, respectively; and m_1, m_2, and m_3 are the masses of water, protein, and alcohol, respectively.

3.5. Effect of Preferential Solvation and Hydration

The enzyme activity and sorption experiments can be summarized as follows. Three composition regimes were found for ethanol (monohydric alcohol):

(i) Lysozyme is in the native (preferentially hydrated) state at high water content. Lysozyme has a higher affinity for water than for alcohols. The ΔG_1^{pref}, ΔG_2^{pref}, and $(\partial g_3/\partial g_2)_{T,\mu 1,\mu 3}$ values are negative. On the other hand, the $(\partial g_1/\partial g_2)_{T,\mu 1,\mu 3}$, Z_1^E, and R_1^E values are positive.

(ii) The preferential binding of ethanol molecules to lysozyme was observed at intermediate water content. The $(\partial g_3/\partial g_2)_{T,\mu 1,\mu 3}$, ΔG_1^{pref}, and ΔG_2^{pref} values are positive. On the other hand, the $(\partial g_1/\partial g_2)_{T,\mu 1,\mu 3}$, Z_1^E, and R_1^E values are negative. Alcohol-induced irreversible inactivation was found in this concentration range (Figure 7).

(iii) The deep dehydration of lysozyme leads to the formation of a rigid (glassy-like) state as a consequence of the proton-transfer phenomena and hydrogen bonding between ionizable and polar protein residues. [1, 2, 7, 37]. The bulky ethanol molecules are incapable of breaking the dehydration-induced protein-protein contacts by themselves, in the absence of water (hydrogen bond donating and hydrogen accepting molecule), likely because of steric and diffusion limitations. The molar volume of ethanol (The molar volume of ethanol is 58.6 cm^3/mol.) is more than three times larger than that of water (18 cm^3/mol). For this reason, the residual enzyme activity is close to ~60% when lysozyme was introduced into water-poor ethanol.

Hence, it is expected that bulky ethanol molecules are not effective in solvating the dehydrated protein alone. This indicates that the ethanol molecules are preferentially excluded from the enzyme surface. It results in the preferential hydration of lysozyme. Therefore, the $(\partial g_1/\partial g_2)_{T,\mu 1,\mu 3}$ values are positive at low water content.

Two distinct schemes are operative for ethylene glycol:

(i) Lysozyme is preferentially hydrated at intermediate and high water content. There is a deficiency of alcohol molecules in the immediate domain of lysozyme. The ΔG_2^{pref} values are negative. The residual enzyme activity values are close to 90-100%.

(ii) In contrast to ethanol, ethylene glycol (polyol) is effective in solvating the dehydrated protein alone, in the absence of water. Therefore, the preferential binding of ethylene glycol to lysozyme was found at low water content. The $(\partial g_3/\partial g_2)_{T,\mu_1,\mu_3}$ ΔG_1^{pref}, and ΔG_2^{pref} values are positive. On the other hand, the ΔG_3^{pref} and $(\partial g_1/\partial g_2)_{T,\mu_1,\mu_3}$ values are negative. Lysozyme is irreversibly inactivated in this concentration region. No residual enzyme activity was observed in the water-poor polyhydric alcohol.

REFERENCES

[1] Schrier, E.E., Ingwall, R.T., Scheraga, H.A. (1965). The effect of aqueous alcohol solutions on the thermal transition of ribonuclease. *J. Phys. Chem.* 69, 298-303.

[2] Parodi, R.M., Bianchi, E., Ciferri, A. (1973). Thermodynamics of unfolding of lysozyme in aqueous alcohol solutions. *J. Biol. Chem.* 248, 4047-4051.

[3] Fujita, Y., Miyanaga, A., Noda, Y. (1979). Effect of alcohols on the thermal denaturation of lysozyme as measured by differential scanning calorimetry. *Bull. Chem. Soc. Japan.* 52, 3659-3662.

[4] Velicelebi, G., Sturtevant, J. (1979). Thermodynamics of the denaturation of lysozyme in alcohol-water mixtures. *Biochem.* 18, 1180-1186.

[5] Brandts, J.F., Hunt, L. (1967). The thermodynamics of protein denaturation. III. The denaturation of ribonuclease in water and in aqueous urea and aqueous ethanol mixtures. *J. Amer. Chem. Soc.* 89, 4826-4838.

[6] Banipal, T.S., Singh, G. (2004). Thermodynamic study of solvation of some amino acids, diglycine and lysozyme in aqueous and mixed aqueous solutions. *Thermochim. Acta.* 412, 63-83.

[7] Timasheff, S. N. (2002). Protein-solvent preferential interactions, protein hydration, and the modulation of biochemical reactions by solvent components. *Proc. Natl. Acad. Sci. USA.*, 99, 9721-9726.

[8] Arakawa, T., Kita, Y. & Timasheff, S. N. (2007). Protein precipitation and denaturation by dimethyl sulfoxide. *Biophys. Chem.*, 131, 62-70.

[9] Gekko, K., Ohmae, E., Kameyama, K. & Takagi, T. (1998). Acetonitrile-protein interactions: Amino acid solubility and preferential solvation. *Biochim. Biophys. Acta*, 1387, 195-205.

[10] Kovrigin, E. L. & Potekhin, S. A. (2000). On stabilizing action of protein denaturants: Acetonitrile effect on stability of lysozyme in aqueous solutions. *Biophys. Chem.*, 83, 45-59.

[11] Shimizu, S. & Matubayasi, N. (2014). Preferential solvation: Dividing surface vs excess numbers. *J. Phys. Chem. B.*, 118, 3922-3930.

[12] Auton, M., Bolen, D. W. & Rosgen, J. (2008). Structural thermodynamics of protein preferential solvation: Osmolyte solvation of proteins, amino acids, and peptides. *Proteins*, 73, 802–813.

[13] Casassa, E. F. & Eisenberg, H. (1964). Thermodynamic analysis of multicomponent solutions. *Adv. Protein Chem.*, 19, 287−395.

[14] Tanford, C. (1969). Extension of theory of linked functions to incorporate effects of protein hydration. *J. Mol. Biol.*, 39, 539−544.

[15] Schellman, J. A. (1987). Selective binding and solvent denaturation. *Biopolymers*, 26, 549−559.

[16] Arakawa, T., Bhat, R. & Timasheff, S. N. (1990). Why preferential hydration does not always stabilize the native structure of globular proteins. *Biochem.*, 29, 1924-1931.

[17] Smith, P. E. (2006). Equilibrium dialysis data and the relationships between preferential interaction parameters in biological systems in terms of Kirkwood-Buff integrals. *J. Phys. Chem. B*, *110*, 2862−2868.

[18] Parsegian, V. A., Rand, R. P. & Rau, D. C. (2000). Osmotic stress, crowding, preferential hydration, and binding: A comparison of perspectives. *Proc. Natl. Acad. Sci. U.S.A.*, *97*, 3987−3992.

[19] Kirby Hade, E. P. & Tanford, C. (1967). Isopiestic compositions as a measure of preferential interactions of macromolecules in two-component solvents. Application to proteins in concentrated aqueous cesium chloride and guanidine hydrochloride. *J. Amer. Chem. Soc.*, *89*, 5034-5040.

[20] Kamiyama, T., Liu, H. L. & Kimura, T. (2009). Preferential solvation of lysozyme by dimethyl sulfoxide in binary solutions of water and dimethyl sulfoxide. *J. Therm. Anal. Cal.*, *95*, 353-359.

[21] Reisler, E., Haik, Y. & Eisenberg, H. (1977). Bovine serum albumin in aqueous guanidine hydrochloride solutions. Preferential and absolute interactions and comparison with other systems. *Biochem.*, *16*, 197-203.

[22] Izumi, T., Yoshimura, Y. & Inoue, H. (1980). Solvation of lysozyme in water/dioxane mixtures studied in the frozen state by NMR spectroscopy. *Arch. Biophys. Biochem.*, *200*, 444-451.

[23] Gregory, R. B. (1995). Protein hydration and glass transition behavior. In: *Protein-Solvent Interactions* (Gregory, R.B., ed.), Marcel Dekker, New York, 191-264.

[24] Rupley, J. A. & Careri, G. (1991). Protein hydration and function. *Adv. Protein Chem.*, *41*, 37-172.

[25] Kuntz, I. D. & Kauzmann, W. (1974). Hydration of proteins and polypeptides. *Adv. Protein Chem.*, *28*, 239-345.

[26] Oleinikova, A., Smolin, N., Brovchenko, I., Geiger, A. & Winter, R. (2005). Properties of spanning water networks at protein surfaces. *J. Phys. Chem. B.*, *109*, 1988-1998.

[27] Durchschlag, H. & Zipper, P. (2001). Comparative investigations of biopolymer hydration by physicochemical and modeling techniques. *Biophys. Chem.*, *93*, 141-157.

[28] Sirotkin, V. A. & Khadiullina, A. V. (2013). Gibbs energies, enthalpies, and entropies of water and lysozyme at the inner edge of excess hydration. *J. Chem. Phys.*, *139*, 075102/1-075102/9.

[29] Sirotkin, V. A. & Khadiullina, A. V. (2014). A study of the hydration of ribonuclease A using densitometry: Effect of the protein hydrophobicity and polarity. *Chem. Phys. Lett*, *603*, 13-17.

[30] Privalov, P. L. & Crane-Robinson, C. (2017). Role of water in the formation of macromolecular structures. *Eur. Biophys. J.*, *46*, 203-224.

[31] Bull, H. B. (1944). Adsorption of water vapor by proteins. *J. Amer. Chem. Soc.*, 66, 1499-1507.

[32] Luscher-Mattli, M. & Ruegg, M. (1982). Thermodynamic functions of biopolymer hydration. I. Their determination by vapor pressure studies, discussed in an analysis of the primary hydration process. *Biopolymers*, 21, 403-418.

[33] Luscher-Mattli, M. & Ruegg, M. (1982). Thermodynamic functions of biopolymer hydration. II. Enthalpy-entropy compensation in hydrophilic hydration process. *Biopolymers*, 21, 419-429.

[34] Bone, S. (1987). Time-domain reflectometry studies of water binding and structural flexibility in chymotrypsin. *Biochem. Biophys. Acta.*, *916*, 128-134.

[35] Sirotkin, V. A. & Khadiullina, A. V. (2011). Hydration of proteins: excess partial enthalpies of water and proteins. *J. Phys. Chem. B*, *115*, 15110-15118.

[36] Sirotkin, V. A., Komissarov, I. A. & Khadiullina, A. V. (2012). Hydration of proteins: excess partial volumes of water and proteins, *J. Phys. Chem. B*, *116*, 4098-4105.

[37] Sirotkin, V. A. (2005). Effect of dioxane on the structure and hydration-dehydration of α-chymotrypsin as measured by FTIR spectroscopy. *Biochim. Biophys. Acta.*, *1750*, 17-29.

[38] Fersht, A. (1999). *Structure and Mechanism in Protein Science: A Guide to Enzyme Catalysis and Protein Folding*; Freeman & Co: New York.

[39] Lehninger, A. L., Nelson, D. L. & Cox, M. M. (1993). *Principles of Biochemistry*; Worth: New York.

[40] Perrin, D. D., Armarego, W. L. F. & Perrin, D. R. (1980). *Purification of Laboratory Chemicals*, Oxford: Pergamon Press.

[41] Sirotkin, V. A. & Kuchierskaya, A. A. (2017). Preferential solvation/hydration of α-chymotrypsin in water-acetonitrile mixtures. *J. Phys. Chem. B.*, *121*, 4422-4430.

[42] Sirotkin, V. A. & Kuchierskaya, A. A. (2017). Lysozyme in water-acetonitrile mixtures: Preferential solvation at the inner edge of excess hydration. *J. Chem. Phys.*, *146*, 215101-8.

[43] Sirotkin, V. A. & Kuchierskaya, A. A. (2017). α-Chymotrypsin in water-ethanol mixtures: Effect of preferential interactions. *Chem. Phys. Lett.*, *689*, 156-161.

[44] Sirotkin, V. A. & Kuchierskaya, A. A. (2017). α-Chymotrypsin in water-acetone and water-dimethyl sulfoxide mixtures: Effect of preferential solvation and hydration. *Proteins: Functions, Structure and Bioinformatics*, *85*, 1808-1819.

[45] Sirotkin, V. A., Zinatullin, A. N., Solomonov, B. N., Faizullin, D. A. & Fedotov, V. D. (2002). Interaction enthalpies of solid bovine pancreatic chymotrypsin with organic solvents: comparison with FTIR-spectroscopic data. *Thermochim. Acta*, *382*, 151-160.

[46] Atkins, P. W. (2006). *Physical Chemistry*. 8th ed. Oxford: Oxford University Press.

[47] Prausnitz, J. M. (1969). *Molecular Thermodynamics of Fluid-Phase Equilibria*. N.J.: Prentice-Hall, Inc., Engelwood Cliffs.

[48] Belousov, V. P. & Panov, M. Y. (1994). *Thermodynamic properties of aqueous solutions of organic substances*. Boca Raton, Fla.: CRC Press.

[49] Bell, G., Janssen, A. E. M. & Halling, P. (1996). Water activity fails to predict critical hydration level for enzyme activity in polar organic solvents: Interconversion of water concentrations and activities. *Enzym. Microb. Technol.*, 20, 471-476.

AUTHOR'S CONTACT INFORMATION

Dr. Vladimir A. Sirotkin, PhD
Kazan Federal University
A.M. Butlerov Institute of Chemistry
Associate Professor
Email: vsir@mail.ru

INDEX

α

α-chymotrypsin, 21, 23, 25, 26, 28, 31, 32, 39, 40, 41, 42, 52, 55, 61, 62, 64, 65, 79, 80, 81, 85, 86, 87, 89, 91, 93, 119, 120, 121, 147

β

β-sheet, 32, 80, 81
β-turns, 32, 78, 81

A

absorbance, 30, 31, 32, 64, 77, 78, 79, 80, 81, 96, 128
acetone, 23, 55, 89, 94, 120, 147
acetonitrile, v, 8, 10, 16, 20, 21, 23, 51, 53, 55, 57, 58, 61, 62, 64, 65, 66, 67, 68, 69, 70, 71, 72, 73, 74, 75, 76, 77, 78, 79, 81, 82, 86, 87, 89, 115, 119, 120, 144, 147
acetonitrile activity coefficients, 74
acid, 21, 33, 53, 87, 115, 144
activity coefficients, 6, 7, 10, 11, 17, 18, 44, 45, 73, 106, 108, 110, 111, 126, 139, 140

alcohol, v, 2, 17, 25, 26, 27, 28, 29, 30, 33, 34, 35, 36, 37, 38, 39, 40, 41, 42, 43, 44, 45, 46, 47, 48, 49, 53, 63, 73, 123, 124, 125, 126, 127, 128, 129, 130, 131, 132, 133, 134, 135, 136, 137, 138, 139, 140, 141, 142, 143
alters, 40, 135
amide I, 31, 32, 33, 57, 77, 78, 79, 80, 81
amide I band, 31, 32, 57, 77, 80
amino, 20, 21, 51, 53, 54, 86, 87, 93, 115, 118, 125, 144
amino acid, 20, 21, 51, 53, 54, 86, 87, 93, 115, 118, 125, 144
amyloid fibril, vii, 26, 58
aqueous solutions, 21, 23, 53, 56, 58, 87, 89, 115, 121, 144, 148
aqueous-organic mixtures, 10, 17, 44, 73, 92, 104, 106, 110, 139
ATEE, 33
atmospheric pressure, 12, 96

B

bilateral, 26, 27, 124
bilateral action, 26, 124
binary mixtures, 1, 3, 4, 5, 6, 11, 12, 111

biocatalysis, vii, 2, 20, 21, 26, 51, 52, 85, 87, 118, 119
biocatalysts, 59, 92
biological systems, 22, 54, 88, 116, 145
biomacromolecule, 60, 92
biopolymer, 49, 50, 84, 117, 146
biotechnological applications, vii, 58, 92
biotechnology, 2, 28
bonding, 4, 48, 82, 93, 142
bonds, 2, 28, 61, 62, 82, 93, 114
bovine pancreatic α-chymotrypsin, 28, 62

C

carbon, 30, 64
carbon dioxide, 30, 64
catalysis, 2, 59
catalytic activity, 21, 38, 52, 69, 70, 86, 100, 101, 119, 123, 133
cesium, 22, 54, 88, 116, 145
chemical, 4, 6, 18, 27, 47, 58, 60, 77, 108, 124, 125, 141
chymotrypsin, 21, 23, 25, 26, 28, 29, 31, 32, 33, 34, 35, 36, 37, 38, 39, 40, 41, 42, 43, 45, 46, 47, 48, 49, 50, 52, 55, 57, 61, 62, 63, 64, 65, 66, 68, 69, 70, 71, 72, 76, 77, 78, 79, 80, 81, 84, 85, 86, 87, 89, 90, 91, 93, 118, 119, 120, 121, 146, 147
compensation, 50, 84, 117, 146
composition, 4, 6, 7, 8, 27, 34, 60, 96, 109, 112, 113, 114, 124, 128, 141
contour, 30, 32, 64, 77, 80
crystals, 20, 51, 86, 119

D

deficiency, 82, 142
dehydration, 48, 50, 85, 118, 142, 147
denaturation, vii, viii, 2, 21, 22, 26, 27, 52, 53, 54, 59, 82, 86, 87, 92, 115, 116, 119, 124, 125, 126, 143, 144

densitometry, 12, 50, 60, 84, 92, 96, 117, 146
density, 11, 12, 96, 111
detection, 30, 64
dialysis, 22, 54, 82, 88, 114, 116, 145
dielectric constant, 52, 85
differential scanning, 53, 143
differential scanning calorimetry, 53, 143
diffusion, 48, 142
dimethyl sulfoxide (DMSO), v, 8, 10, 12, 21, 22, 23, 52, 53, 54, 55, 86, 87, 88, 89, 91, 93, 94, 95, 96, 97, 98, 99, 101, 102, 103, 104, 105, 106, 107, 108, 109, 112, 113, 114, 115, 116, 119, 120, 121, 144, 145, 147
distilled water, 13, 96
donors, 62, 93

E

egg, 52, 85, 91, 93, 94, 123, 125, 126
energy, 46, 47, 74, 75, 113
entropy, 4, 50, 84, 117, 146
enzymatic reaction, 34, 38, 39, 69, 70, 96, 100, 101, 128, 133
enzyme, 2, 13, 20, 23, 27, 28, 33, 34, 38, 39, 40, 42, 48, 50, 51, 55, 56, 58, 59, 60, 65, 66, 68, 69, 71, 81, 82, 85, 86, 88, 89, 91, 92, 93, 95, 97, 100, 102, 114, 118, 120, 121, 123, 125, 128, 133, 134, 135, 136, 141, 142, 143, 147, 148
enzyme activity, 23, 28, 33, 40, 48, 56, 58, 71, 81, 82, 89, 91, 95, 102, 114, 121, 123, 125, 128, 135, 141, 148
equilibrium, 7, 17, 19, 29, 44, 63, 73, 82, 95, 106, 108, 114, 127, 139
ester, 28, 33, 62, 76
ethanol, vi, 23, 26, 28, 29, 33, 34, 35, 36, 37, 38, 39, 40, 41, 42, 43, 45, 46, 47, 49, 52, 53, 55, 85, 89, 96, 120, 123, 124, 125, 126, 128, 129, 130, 132, 133, 134,

Index

135, 136, 137, 138, 140, 141, 142, 143, 147
ethylene, 8, 9, 19, 29, 63, 95, 123, 124, 125, 126, 127, 133, 136, 142, 143
ethylene glycol, vi, 8, 9, 19, 29, 63, 95, 123, 124, 125, 126, 127, 132, 133, 135, 136, 142, 143
excess chemical potentials, 6, 18, 47, 77, 108, 141
excess functions, 4, 13, 34, 35, 66, 97, 98, 128, 129
excess partial quantities, 4, 5
exclusion, 60, 125
experimental condition, 30, 64

F

films, 30, 64
flexibility, 50, 59, 84, 118, 146
fluorescence, 21, 52, 86, 119
formamide, 7, 9
formation, 33, 48, 50, 52, 58, 84, 85, 117, 142, 146
FTIR (Fourier Transform Infrared) spectroscopy, v, viii, 30, 50, 57, 61, 63, 64, 85, 90, 118, 120, 147
FTIR spectra, 30, 63
FTIR spectroscopic measurements, 30, 64
FTIR spectroscopy, v, 30, 50, 57, 85, 118, 147
function of mixing, 4, 5, 13, 34, 66, 98, 128
functions, 4, 8, 15, 22, 23, 26, 36, 50, 54, 55, 60, 65, 67, 84, 87, 89, 92, 93, 96, 99, 109, 115, 117, 120, 125, 128, 131, 144, 146, 147

G

Gibbs energies, v, viii, 1, 3, 4, 6, 8, 12, 16, 25, 28, 43, 49, 60, 61, 72, 84, 91, 93, 104, 109, 112, 117, 121, 123, 125, 137, 146
Gibbs energy, 3, 45, 46, 47, 74, 75, 76, 105, 107, 113, 138, 140, 141
Gibbs-Duhem equation, 18, 47, 76, 107, 140
glass transition, 49, 83, 116, 145
glassy (rigid) state, 30, 48, 57, 64, 82, 142
glycerol, 7, 9, 10, 20, 51, 86, 118
glycol, 8, 9, 19, 29, 63, 95, 123, 124, 125, 126, 127, 129, 130, 133, 134, 136, 137, 138, 140, 141, 142, 143
guidelines, 18, 29, 62, 63, 94, 95, 126, 127, 128

H

hen egg-white lysozyme, 91, 93, 94, 125, 126
human, 20, 51, 52, 86, 119, 120
humidity, 30, 64
hydration, v, viii, 6, 7, 8, 9, 10, 20, 23, 25, 27, 28, 40, 49, 50, 51, 52, 55, 56, 57, 58, 59, 60, 61, 62, 63, 81, 83, 84, 85, 86, 89, 93, 95, 97, 102, 108, 109, 114, 116, 117, 118, 119, 120, 121, 123, 124, 125, 135, 145, 146, 147, 148
hydration excess, 7, 8, 9, 10, 109
hydrogen, 2, 4, 48, 62, 82, 93, 113, 114, 115, 142
hydrogen bond accepting ability, 93, 115
hydrogen bonding, 4, 48, 82, 93, 142
hydrogen bonds, 2, 62, 82, 93, 114
hydrolysis, 20, 33, 38, 51, 69, 76, 86, 101, 118, 133
hydrolytic enzymes, 93
hydrolytic reactions, vii, 26, 93
hydrophilic structure breaker, 8
hydrophilic structure maker, 8
hydrophobic, 2, 27, 28, 58, 59, 60, 82, 92, 125
hydrophobicity, 50, 84, 117, 146

hydroxyl, 123, 126
hydroxyl group, vi, 123, 126
hysteresis, 20, 51, 86, 119

I

ideal, 4, 5, 13, 14, 15, 34, 36, 37, 40, 42, 65, 67, 68, 71, 97, 99, 100, 102, 128, 130, 131, 135, 136
ideal binary mixture, 5, 13, 34, 40, 42, 66, 71, 98, 102, 128, 135
ideal system, 4
infrared spectra, 30, 31, 57, 64, 77, 80
inhibition, 55, 89
irreversible enzyme inactivation, 136
isothermal calorimetry, 20, 51, 86, 119
isotherms, 34, 35, 64, 65, 96, 128, 129

K

Karl Fischer titration, 19, 29, 30, 62, 63, 94, 95, 127
kinetic curves, 38, 39, 69, 70, 100, 101, 133

L

liquids, vii, viii, 1, 2, 12, 28, 59, 61, 91, 92, 93, 96, 126
lysozyme, v, vi, 21, 22, 23, 25, 29, 30, 37, 43, 49, 52, 53, 54, 55, 69, 84, 85, 87, 88, 89, 91, 93, 94, 95, 96, 97, 98, 99, 100, 101, 102, 103, 104, 105, 107, 114, 115, 116, 117, 120, 123, 124, 125, 126, 127, 128, 129, 130, 131, 132, 133, 134, 135, 136, 137, 138,140, 141, 142, 143, 144, 145, 146, 147

M

macromolecules, vii, viii, 2, 22, 26, 27, 54, 57, 58, 60, 88, 91, 92, 93, 116, 125, 145
mass, 10, 11, 12, 14, 15, 16, 17, 18, 19, 29, 35, 36, 37, 38, 39, 40, 41, 43, 44, 45, 46, 63, 66, 67, 68, 69, 70, 71, 72, 73, 74, 75, 80, 81, 95, 98, 99, 100, 101, 102, 103, 104, 105, 106, 107, 110, 111, 127, 130, 131, 132, 133, 135, 137, 138, 139, 140
measurements, viii, 12, 19, 30, 33, 60, 61, 64, 96, 101, 134
media, 2, 20, 21, 51, 52, 85, 86, 87, 93, 118, 119
methanol, 26, 28, 32, 33, 34, 35, 36, 38, 39, 40, 41, 42, 43, 45, 46, 47, 49
methodology, v, vii, 1, 2, 3, 6, 13, 25, 56, 90, 112
mixing, 4, 5, 7, 13, 34, 35, 66, 67, 92, 98, 113, 128, 130
molar volume, 48, 49, 142
molecular mass, 28, 62
molecular weight, 94, 126
molecules, viii, 2, 6, 7, 8, 27, 29, 42, 48, 49, 57, 59, 61, 63, 82, 92, 95, 108, 109, 113, 123, 124, 127, 142
molten globule, vii, 26, 52, 58, 85
monohydric alcohols, 25, 26, 27, 48, 49, 124

N

N-acetyl-L-tyrosine ethyl ester, 33, 76
native, 22, 48, 49, 54, 59, 77, 81, 88, 114, 116, 126, 141, 144
neurodegenerative diseases, vii, 26
nonaqueous media, 2, 20, 51, 85, 93, 118

Index   155

O

organic solvent, v, vii, viii, 1, 2, 3, 7, 10, 11, 12, 13, 14, 15, 16, 17, 18, 19, 20, 21, 23, 27, 28, 30, 31, 32, 33, 34, 35, 43, 50, 51, 52, 56, 57, 58, 59, 60, 61, 62, 65, 66, 67, 68, 71, 72, 73, 77, 82, 85, 86, 87, 89, 90, 91, 92, 93, 95, 96, 97, 98, 99, 100, 102, 103, 106, 107, 108, 110, 111, 118, 119, 120, 121, 124, 125, 128, 129, 130, 137, 147, 148

P

partial quantities, 3, 6
peptide, 21, 27, 28, 31, 54, 58, 59, 61, 77, 80, 87, 93, 115, 125, 144
pH, 33, 96, 128
phosphate, 96, 128
polar, 20, 23, 48, 51, 56, 59, 82, 85, 89, 118, 121, 142, 148
polarity, 2, 50, 84, 117, 146
polyhydric alcohols, 125
polyols, 125
polypeptides, 49, 83, 117, 145
polysaccharide, 93, 125
polysaccharide chains, 93, 125
potassium, 33, 96, 128
precipitation, 21, 53, 87, 115, 144
preferential binding, viii, 16, 27, 43, 48, 57, 59, 60, 72, 82, 103, 114, 124, 136, 142, 143
preferential hydration, v, vii, viii, 2, 16, 22, 26, 27, 43, 48, 54, 57, 58, 59, 60, 61, 69, 72, 82, 83, 88, 91, 92, 93, 103, 104, 116, 123, 124, 125, 137, 142, 144, 145
preferential interaction parameters, viii, 1, 16, 22, 42, 43, 46, 54, 61, 72, 74, 75, 88, 103, 104, 107, 116, 136, 137, 140, 145
preferential interactions, 1, 6, 8, 13, 14, 16, 18, 21, 22, 23, 28, 34, 35, 42, 47, 48, 53, 54, 55, 57, 60, 65, 66, 67, 71, 72, 76, 81, 87, 88, 89, 91, 92, 98, 99, 103, 107, 109, 115, 116, 120, 126, 128, 129, 130, 136, 140, 144, 145, 147
preferential solvation, v, vi, viii, 1, 2, 6, 7, 21, 22, 23, 25, 26, 27, 42, 46, 49, 53, 54, 55, 58, 59, 60, 61, 69, 72, 87, 88, 89, 91, 92, 93, 97, 105, 108, 115, 116, 120, 121, 123, 124, 125, 136, 138, 141, 144, 145, 147
preferential solvation and hydration, vi, 1, 23, 55, 89, 120, 123, 141, 147
preferential solvation parameters, 42, 46, 72, 136, 138
propanol-1, 26, 28, 32, 34, 35, 36, 37, 38, 39, 40, 41, 42, 43, 45, 46, 47, 48
protein, vii, viii, 1, 2, 3, 8, 11, 12, 13, 14, 15, 16, 18, 19, 20, 21, 22, 25, 26, 27, 28, 29, 30, 31, 32, 34, 35, 42, 43, 46, 47, 48, 49, 50, 51, 52, 53, 54, 55, 56, 57, 58, 59, 60, 61, 62, 63, 64, 65, 66, 67, 68, 72, 74, 75, 77, 80, 82, 83, 84, 85, 86, 87, 88, 90, 91, 92, 93, 94, 95, 97, 98, 99, 103, 108, 111, 114, 115, 116, 117, 118, 120, 123, 124, 125, 126, 127, 129, 130, 136, 137, 141, 142, 143, 144, 145, 146, 147
protein denaturation, 2, 27, 53, 82, 92, 124, 125, 143
protein destabilization, viii, 25, 59, 91, 92
protein destabilizer, 8
protein folding, 20, 51, 86, 118
protein hydration, vii, 16, 21, 22, 43, 53, 54, 60, 72, 87, 91, 92, 93, 103, 115, 137, 144
protein macromolecule, vii, viii, 2, 26, 27, 57, 59, 60, 91, 92, 93, 125
protein macromolecules, vii, viii, 2, 26, 27, 57, 60, 91, 92, 93, 125
protein stabilization/destabilization, 18, 57
protein stabilizer, 8
protein structure, viii, 59, 61

pure water, 5, 10, 11, 14, 17, 31, 36, 38, 40, 44, 67, 69, 70, 73, 77, 78, 79, 80, 81, 99, 101, 106, 110, 111, 130, 133, 139
purity, 29, 62, 94, 126

R

reactions, vii, 2, 21, 26, 53, 58, 59, 76, 83, 87, 92, 93, 115, 144
reagents, 2, 59
reliability, 47, 74
residual enzyme activity, 13, 27, 33, 34, 38, 39, 40, 42, 66, 68, 69, 71, 91, 93, 95, 97, 100, 102, 114, 123, 125, 128, 133, 134, 135, 142, 143
residues, 48, 93, 125, 142
resolution, 30, 63
room temperature, 82, 114

S

selectivity, 2, 81
serum, 20, 22, 51, 52, 55, 56, 86, 88, 90, 116, 119, 120, 145
serum albumin, 20, 22, 51, 52, 55, 56, 86, 88, 90, 116, 119, 120, 145
solubility, 2, 21, 53, 59, 87, 92, 115, 144
solution, 7, 30, 33, 34, 52, 64, 85, 96, 108, 113, 114, 126, 128
solvation, vi, vii, viii, 1, 2, 6, 7, 8, 11, 12, 13, 14, 15, 16, 17, 18, 21, 22, 23, 25, 26, 27, 34, 35, 36, 37, 38, 41, 42, 43, 45, 46, 49, 53, 54, 55, 58, 59, 60, 61, 65, 66, 67, 68, 69, 71, 72, 74, 75, 87, 88, 89, 91, 92, 93, 97, 98, 99, 100, 102, 103, 104, 105, 106, 107, 108, 109, 110, 111, 112, 113, 115, 116, 120, 121, 123, 124, 125, 128, 129, 130, 131, 132, 133, 135, 136, 137, 138, 139, 140, 141, 144, 145, 147
solvation excess, 7, 108
solvation layer, 7, 8, 11, 12, 13, 14, 15, 16, 17, 18, 27, 34, 35, 36, 37, 38, 41, 42, 43, 45, 65, 66, 67, 68, 71, 72, 74, 92, 98, 99, 100, 102, 108, 109, 110, 111, 112, 113, 124, 128, 129, 130, 131, 132, 133, 136, 137, 139, 140
solvation shell, viii, 2, 7, 8, 15, 37, 42, 45, 46, 59, 68, 71, 75, 97, 98, 99, 100, 102, 103, 104, 105, 106, 107, 108, 109, 112, 113, 131, 132, 135, 138, 140
solvent molecules, viii, 2, 59
solvents, vii, 2, 22, 54, 58, 61, 88, 114, 116, 118, 145
sorption, viii, 1, 2, 7, 13, 15, 18, 19, 20, 25, 27, 29, 30, 34, 35, 36, 37, 47, 48, 51, 52, 57, 61, 63, 64, 65, 66, 67, 74, 81, 86, 91, 93, 95, 96, 97, 99, 114, 119, 123, 125, 127, 128, 129, 131, 132, 141
sorption equilibrium, 19, 29, 63, 95, 127
sorption experiments, 48, 141
sorption isotherms, 34, 35, 64, 65, 96, 128, 129
sorption measurements, 18, 19, 29, 63, 95, 127
spectral experiments, 81
spectroscopy, viii, 22, 50, 55, 60, 61, 85, 88, 92, 116, 118, 145, 147
stability, vii, 21, 26, 27, 49, 53, 58, 82, 87, 92, 113, 115, 124, 125, 144
stabilization, viii, 2, 18, 25, 47, 57, 59, 76, 91, 92, 107, 125, 126, 140
stabilizers, 55, 89
standard error, 97, 129
state, 10, 17, 22, 44, 46, 47, 48, 55, 57, 59, 60, 73, 75, 76, 77, 80, 81, 82, 88, 106, 107, 110, 114, 116, 139, 141, 142, 145
states, vii, 26, 52, 58, 85
storage, 33, 95, 128
stress, 22, 54, 56, 88, 90, 116, 145
stress factors, 56, 90
structural characteristics, 30, 64

structure, vii, viii, 2, 8, 21, 22, 23, 26, 31, 32, 33, 50, 52, 54, 55, 56, 57, 58, 59, 61, 62, 77, 80, 81, 82, 85, 87, 88, 89, 90, 116, 118, 119, 120, 125, 126, 144, 147
substrate, 33, 34, 92, 96, 128
suppression, vii, 2, 26, 59, 65, 69
suspensions, 52, 86, 119
synthesis, 2, 58, 59, 93
synthetic reactions, 59, 83

T

techniques, 27, 49, 84, 92, 117, 124, 146
temperature, 3, 4, 19, 27, 29, 30, 63, 64, 95, 124, 125, 126, 127
ternary systems, 18, 47, 76, 107, 140
thermal stability, vii, 26, 82
thermodynamic functions, 1, 3, 60
thermodynamic properties, 4, 6
thermodynamics, 21, 53, 54, 87, 115, 143, 144
thermostability, vii, 2, 26, 59, 114
transesterification, vii, 26, 59, 76, 83, 93
trypsin, 21, 52, 87, 119
tryptophan, 21, 52, 86, 119
tyrosine, 28, 33, 62, 76

U

urea, vii, 27, 53, 92, 125, 143

V

vapor, 7, 17, 18, 20, 29, 34, 44, 50, 51, 60, 63, 64, 73, 84, 86, 92, 95, 96, 106, 108, 117, 119, 127, 128, 139, 146
volumes, 4, 5, 50, 60, 84, 118, 121, 147

W

water activity, 7, 10, 11, 17, 23, 44, 45, 56, 73, 74, 89, 106, 108, 110, 121, 139, 148
water activity coefficients, 10, 11, 17, 44, 45, 73, 74, 106, 110, 126, 139
water binding, 125
water mass fraction, 12, 14, 15, 16, 35, 36, 37, 38, 40, 41, 43, 46, 67, 68, 69, 71, 72, 75, 80, 81, 98, 99, 100, 102, 103, 104, 105, 111, 130, 131, 132, 133, 135, 137, 138
water sorption, 13, 19, 30, 34, 37, 47, 51, 52, 63, 65, 66, 74, 86, 97, 119, 125, 127, 128, 129, 132
water vapor, 30, 50, 64, 84, 117, 146
water-alcohol mixture, 25, 27, 29, 34, 35, 36, 37, 38, 39, 40, 41, 42, 43, 47, 124, 127, 128, 129, 130, 131, 133, 134, 135, 136, 137, 141
water-miscible organic solvent, 93, 125
water-organic mixture, 7, 8, 13, 14, 15, 16, 19, 29, 35, 37, 40, 45, 46, 57, 61, 62, 63, 66, 67, 68, 69, 71, 72, 75, 94, 95, 96, 98, 99, 100, 102, 103, 105, 108, 109, 115, 126, 128, 130, 132, 135, 138, 140
water-organic mixtures, 7, 8, 13, 14, 16, 29, 35, 40, 45, 46, 57, 61, 62, 67, 71, 72, 75, 94, 95, 96, 98, 100, 102, 103, 105, 108, 109, 115, 126, 128, 130, 135, 138, 140
water-poor, vii, 26, 34, 40, 59, 92, 96, 123, 128, 135, 136, 142, 143
water-protein interactions, 92

Z

zero hydration level, 29, 63, 95, 127

Related Nova Publications

FURAN: CHEMISTRY, SYNTHESIS AND SAFETY

EDITOR: Ida Bailey

SERIES: Chemistry Research and Applications

BOOK DESCRIPTION: In this compilation, the authors discuss the reactions of furans and formylfurans with aqueous hydrogen peroxide which are caused by peculiarities in the chemical behavior of these reagents, as well as their availability.

SOFTCOVER ISBN: 978-1-53615-390-3
RETAIL PRICE: $82

PHOTOISOMERIZATION: CAUSES, BEHAVIOR AND EFFECTS

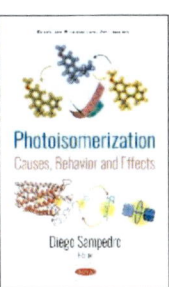

EDITOR: Diego Sampedro

SERIES: Chemistry Research and Applications

BOOK DESCRIPTION: New and exciting applications of light-controlled processes have become practical in the diagnosis and treatment of diseases, the preparation and use of functional materials, the storage of solar energy and the control of biological properties. Many of these applications are based on a very simple chemical step: a photoisomerization.

HARDCOVER ISBN: 978-1-53615-313-2
RETAIL PRICE: $195

To see a complete list of Nova publications, please visit our website at www.novapublishers.com

Related Nova Publications

A Closer Look at Calorimetry

Editor: Oliver Wrigley

Series: Chemistry Research and Applications

Book Description: In this compilation, the authors review the planning of multithermal titration calorimetry experiments using triosephosphate isomerase as a case study with two of its inhibitors, 2PG and PGH, under physiological conditions and osmotic stress.

Softcover ISBN: 978-1-53615-789-5
Retail Price: $82

Thiols: Structure, Properties and Reactions

Editor: Carlos C. McAlpine

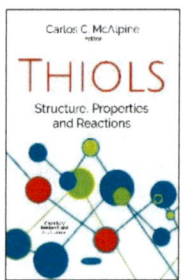

Series: Chemistry Research and Applications

Book Description: The thiol (-SH) is an essential functional group in the biological system. In this compilation, the authors begin by exploring how the inorganic and organic forms of mercury interact with these macromolecules.

Softcover ISBN: 978-1-53615-599-0
Retail Price: $82

To see a complete list of Nova publications, please visit our website at www.novapublishers.com